THE ANTI-ALZHEIMER'S PRESCRIPTION

The Science-Proven Plan to Start at Any Age

VINCENT FORTANASCE, M.D.

GOTHAM BOOKS

Neither the publisher nor the author is engaged in rendering professional advice or services to the individual reader. The ideas, procedures, and suggestions contained in this book are not intended as a substitute for consulting with your physician. All matters regarding your health require medical supervision. Neither the author nor the publisher shall be liable or responsible for any loss or damage allegedly arising from any information or suggestion in this book.

GOTHAM BOOKS
Published by Penguin Group (USA) Inc.
375 Hudson Street, New York, New York 10014, U.S.A.
Penguin Group (Canada), 90 Eglinton Avenue East, Suite 700, Toronto,
Ontario M4P 2Y3, Canada (a division of Pearson Penguin Canada Inc.);
Penguin Books Ltd, 80 Strand, London WC2R 0RL, England;
Penguin Ireland, 25 St Stephen's Green, Dublin 2,
Ireland (a division of Penguin Books Ltd);
Penguin Group (Australia), 250 Camberwell Road, Camberwell,
Victoria 3124, Australia (a division of Pearson Australia Group Pty Ltd);
Penguin Books India Pvt Ltd, 11 Community Centre,
Panchsheel Park, New Delhi—110 017, India;
Penguin Group (NZ), 67 Apollo Drive, Rosedale, North Shore 0632,
New Zealand (a division of Pearson New Zealand Ltd);
Penguin Books (South Africa) (Pty) Ltd, 24 Sturdee Avenue,
Rosebank, Johannesburg 2196, South Africa

Penguin Books Ltd, Registered Offices: 80 Strand, London WC2R 0RL, England

Published by Gotham Books, a division of Penguin Group (USA) Inc.

First printing, August 2008
5   7   9   10   8   6   4

Gotham Books and the skyscraper logo are trademarks of Penguin Group (USA) Inc.

LIBRARY OF CONGRESS CATALOGING-IN-PUBLICATION DATA
Fortanasce, Vincent.
The anti-Alzheimer's prescription : the science-proven plan to start at any age / Vincent Fortanasce.
p.   cm.
Includes bibliographical references.
ISBN 978-1-592-40379-0 (hardcover)   1. Alzheimer's disease—Popular works.
2. Alzheimer's disease—Prevention. I. Title.
RC523.F6883 2008
616.8'3105—dc22   2007046910

Printed in the United States of America
Set in Adobe Garamond
Designed by Elke Sigal

While the author has made every effort to provide accurate telephone numbers and Internet addresses at the time of publication, neither the publisher nor the author assumes any responsibility for errors, or for changes that occur after publication. Further, the publisher does not have any control over and does not assume any responsibility for author or third-party Web sites or their content.

To my father, my "Ole Sox," my orchestra, and the man who was always there for us. As a father, I now realize the incredible gift he gave to me. Dad, for the good times we spent at Ebbets Field when I was a boy; for the difficult times in my training; for your faith in me as a young man; for the times I still feel your presence in my life today; for the love and respect you taught us in the way that you loved our mother . . . thank you.

To Joe Van Der Meulen, M.D., my professor, my mentor, my conductor, the man who taught me what it was like to be a physician. Your words, "Patients don't care how much you know until they know how much you care," changed my life. Like Pope John Paul II, there is no one who worked with you or whom you taught who doesn't feel blessed to have you in their lives.

To Joan Lasorda, one of the unheralded women in America. There is a reason why Tommy is so ebullient and radiates goodness; it is because of you, Joan. The relationship of mutual respect and love you and Tommy have for each other reflects the real foundation of our society. I dedicate this book to you and all women like you, including my mother and sisters, who through their altruism give life harmony.

CONTENTS

Foreword   *ix*

Preface   *xi*

PART I
UNDERSTANDING ALZHEIMER'S DISEASE

1   Is Alzheimer's in Your Future?   *3*

2   Alzheimer's Disease: Are You at Risk?   *29*

3   Test Your Real Brain Age   *57*

PART II
THE 4-STEP ANTI-ALZHEIMER'S PRESCRIPTION

4   Step 1: The Anti-Alzheimer's Diet: Food for Thought   *75*

5   Step 2: Brawn Boosters: Daily Aerobics and
Anaerobics for the Body and Mind   *112*

6   Step 3: Brain Boosters:
Daily Neurobics to Build a Big Brain's Reserve   *146*

7   Step 4: Rest and Recovery: Finding Your Circle of Quiet   *174*

PART III
DIAGNOSIS, TREATMENT, AND
THE FUTURE OF ALZHEIMER'S DISEASE

8   Is It Alzheimer's? Making the Diagnosis   *207*

9   The Latest Medical Therapies for Alzheimer's Disease   *231*

Appendix A:
Anti-Alzheimer's Diet Shopping List,
28-Day Menu, and Recipes   *253*

Appendix B:
Figure Out Your Body Mass Index   *303*

Appendix C:
Strength Fitness Test   *307*

References and Supporting Research   *311*

Acknowledgments   *317*

Index   *323*

FOREWORD

have known Dr. Vincent Fortanasce for thirty-three years, first as a diligent resident in neurology, then as a well-respected colleague. It is with pleasure that I write the foreword to his latest book, *The Anti-Alzheimer's Prescription*. In it, Vince has developed a novel and unique approach to delaying or preventing the effects of Alzheimer's disease. If you are concerned about your personal risk of Alzheimer's or the mental health of someone you love, it could make a difference.

Driven by the emotional experience of dealing with the ravages of Alzheimer's disease in his own father and the problems he saw in patients in his own neurological practice, he reviewed the literature and drew from his experience to develop the Anti-Alzheimer's Disease Prescription.

Few authors, if any, who have written on this subject have the broad training in neurology, psychiatry, and rehabilitation that Dr. Fortanasce has. This enables him to bring his skills to his integrated Prescription for Alzheimer's Disease that works toward achieving a healthful balance of brain, mind, and body functions. His narrative emphasizes how personal experience can give a whole different emotional aspect to a clinician's impersonal impression of a disease, making the book heartwarming and personal.

Dr. Fortanasce explains his concepts in a step-by-step easily under-standable style for those not familiar with medical terminology and uses numerous illustrations from his private practice. He also introduces some of the more recent concepts of the role inflammation and apopto-sis, or programmed cell death, may have in the etiology of the disease.

His approach to "sentinel" risk factors—sleep and stress—is novel. Although Dr. Fortanasce emphasizes their role in Alzheimer's disease, it is apparent that his recommendations on balancing forces in daily living with exercise, diet, reduction of stress, meditation, and avoidance of risk factors could serve as a template for anyone interested in living a long, healthy, and productive life.

Dr. Fortanasce's dedication to his patients and medical ethics has been recognized nationally. He has been awarded certificates of commendation from the White House, President Clinton, Governor Pete Wilson, United States Senators Barbara Boxer and Dianne Feinstein; and received International Little League's highest honor, being inducted into the Little League Hall of Excellence with other notables such as President Bush and U.S. Senator William Bradley, for his contribution to children's health and sports.

—*Joseph P. Van Der Meulen, M.D.*

Joseph P. Van Der Meulen, M.D., is a past chairman of the department of neurology at the University of Southern California and Los Angeles County Hospital. He was also vice president for Health Affairs at USC, interim dean of the School of Medicine, and Dr. Fortanasce's mentor.

PREFACE

The fact that you are reading this book suggests you are concerned about your memory and your risk of Alzheimer's disease. You are not alone, especially if you're a baby boomer rapidly approaching retirement years, or you're over sixty-five and seeing your contemporaries or even your spouse declining before your eyes.

I wrote this book because I know that Alzheimer's disease can be prevented for most people and delayed in those who are genetically predisposed. I'm interested not only in the health of your brain, but also of your body and spirit. I want to show you how taking care of yourself today will keep your brain sharp and you independent for the rest of your life. I want you to recognize the harmful behaviors (risk factors) that are a threat to your brain, causing it to prematurely age, and how to change those behaviors. I want you to learn the four steps you can take to decrease the chances that you'll get Alzheimer's disease, including dietary changes, exercises, brain boosters, and relaxation techniques.

The good news is that many Americans reaching sixty-five today have an average life expectancy of an additional two decades. By the year 2050, an estimated 40 percent of sixty-five-year-olds are likely to reach age ninety. Now, here's the bad news: Cognitive decline begins as early as age twenty-four, and the pathological changes in the brain associated with Alzheimer's, including the elevated levels of beta-amyloid peptides that cause plaque buildup in the brain (the main characteristic of Alzheimer's), may begin more than three decades before the first signs of the disease are noticeable. At the peak of your brain's abilities, you may be seeding its rapid decline. The incidence of Alzheimer's in people less than sixty-five is growing at an alarming rate.

Modern medicine has overwhelmingly succeeded in increasing the longevity of our bodies. Currently, even the latest medical breakthroughs have failed to increase the longevity of our brains or to reverse their decline. Despite the hype in the news media, the American Academy of Neurology admits that there is not one drug or treatment to stop brain aging, particularly Alzheimer's disease, despite the billions of dollars that are being spent to develop new medications. In the United States today, 50 percent of adults at eighty-five have dementia and it is twice as prevalent in women as in men.

So, what's the point of increased life span if you no longer have a healthy mind, especially if you're spending your last years alone and forgotten in a nursing home, as are 70 percent of those diagnosed with Alzheimer's? That's where my four-step program of diet, exercise, mental aerobics, and rest and relaxation comes into play—as the first scientifically substantiated, multifaceted game plan developed by a neurologist *to stop brain aging*, particularly Alzheimer's disease.

I am board certified in neurology and in rehabilitation medicine and trained in psychiatry at the Institute of Living, a Yale affiliate hospital. Over the past three decades, I have diagnosed and treated thousands of men and women with Alzheimer's disease. Many patients came to my clinic after a concerned family member or friend noticed obvious signs and symptoms of Alzheimer's, such as memory loss, difficulty with speaking, reasoning, or learning, personality and behavioral changes, and problems with activities of daily living such as bowel and bladder incontinence. Others came because they have a strong family history of Alzheimer's. While these patients are asymptomatic now, they have witnessed the horror of Alzheimer's disease in a parent or grandparent and are worried about enduring the same genetic fate. In 1980, I diagnosed, on average, one patient a month with dementia; now, in some weeks, I may diagnose six to ten patients. I truly believe that this dramatic increase in Alzheimer's disease is a direct result of our frenetic, sedentary, fast-food lifestyle.

Since I started my neurology practice, I have seen the countless ways in which Alzheimer's disease can destroy the lives of older adults. Nevertheless, Alzheimer's is not just about the devastation of the patient's mind and life—it shakes the foundation of the entire family and community. Most scientists doubt any cure for Alzheimer's is on the horizon. The only rational approach that physicians agree upon is preventing this mind destroyer with specific lifestyle changes. I wrote this book to show you how taking four steps for self-care today will help you to have a healthier, Alzheimer's-free brain for many years to come.

Throughout this book, I will help you recognize specific risk factors that increase your chances of getting Alzheimer's disease and teach you ways to control or reduce these risk factors. While many physicians still believe that it's pretty much set in stone whether or not you'll get Alzheimer's disease, *I completely disagree!* I know that if you work on changing the risk factors you can control, you can definitely reduce the chance of Alzheimer's disease—even with aging and even with a family history of Alzheimer's.

In addition, I'm going to tell you how to help your children and grandchildren prevent Alzheimer's disease by making brain-healthy diet and exercise choices early in life. It is my firm belief that the road to Alzheimer's begins in childhood. The habits and routines we establish for our children and grandchildren are reflected and perpetuated throughout their lives.

Above all, I want to give you hope—particularly if you have a genetic risk of Alzheimer's disease as I do. By learning more about Alzheimer's disease, you can start today to protect the health of your brain. To that end, the four-step program will provide you with the latest information and specific tools you need to determine a well-planned blueprint to building a bigger and better brain.

—*Vince Fortanasce, M.D.*
  October 15, 2007

THE
ANTI-ALZHEIMER'S
PRESCRIPTION

UNDERSTANDING ALZHEIMER'S DISEASE

1

IS ALZHEIMER'S IN YOUR FUTURE?

The greatest fear for baby boomers is not death, but Alzheimer's disease. Until recently, almost all of the scientific studies and clinical trials concerning Alzheimer's disease were focused on finding powerful medications to help manage the cognitive decline and memory loss, so adults with early Alzheimer's could have a few months to get their affairs in order before dementia robbed them of their thoughts, personality, and dignity. Well aware of the highly advertised drugs that "claim" to boost cognitive function with Alzheimer's disease, the general population and media have received too little information on specific ways to prevent Alzheimer's altogether. That's the reason I wrote this book.

As a neurologist and Alzheimer's specialist, I am passionate about telling people how to prevent or delay Alzheimer's disease. Each day I make a point to talk to as many people as I can about self-care strategies, whether to my patients and their families, my medical students at USC, or my worldwide radio show audience. This strong desire to educate others about preventing Alzheimer's is triggered by years of research and training as a psychiatrist at the Yale-affiliated Institute of Living, as a neurologist at USC, and as a board-certified rehabilitation specialist, along with my personal genetic predisposition. You see, my father was just a few years older than I am right now when he had his first symptoms of Alzheimer's disease.

I still remember that moment, as if it were yesterday. I had rushed from my rehab clinic to the Los Angeles International Airport that January afternoon to pick up my parents for their annual winter reprieve. No matter what your age, a hug from your mom and dad is welcomed and nurturing. Yet this time when I hugged my dad, it was different: He felt distant, like an empty shell without warmth and feeling. Admittedly, I had been concerned about my father for a while. But until now, I had reasoned that he was feeling low because of the recent deaths of two close friends.

Warm, compassionate, and highly spirited, my father was my lifelong mentor, my confidant, and even my Little League coach, as we shared a love for baseball. In my first year of medical school, when I sometimes found myself overwhelmed, my dad took the time to write me weekly letters (before there was e-mail), describing the family's activities and the latest scoop in the baseball world. In four years, Dad never missed a week. He'd start each letter with "Vinnie Ole Sox," an endearment he'd used for as long as I can remember, and he'd always sign each letter with, "Your Daddy Ole Sox." Along with the lively update of the family, Dad always assured me of his confidence and faith in me and my goal to become a neurologist, a doctor who specializes in diseases of the brain.

However, that sunny afternoon as I carried my parents' luggage to the car, I noticed that my dad was unusually quiet and withdrawn. I asked him about the flight to break the silence. "Did you have a lot of turbulence this time, Dad?"

Before he could respond, my mom quickly replied, "We thought the trip was just fine."

I tried again. "Dad, how was lunch? I hope the food was better this trip."

Again, my father said nothing while my mother quickly interjected, "The pasta was like mush. Your dad would not eat it."

That night at dinner, Dad was abnormally quiet. Later, as we watched television with the children, I noticed that my father confused our children's names. My mother also noticed and quickly covered for him. "Ah! Vin, they are growing so fast that your father and I can hardly recognize them anymore!"

The next evening when I returned home from the hospital, my dad was sitting in the dining room alone, staring blankly out the window.

Concerned, I put my hand on his shoulder and asked, "Daddy Ole Sox, how ya doing?"

Dad didn't reach out and take my hand, as was his custom. Instead, his response was an emotionless, "Okay."

Later, I asked my wife if she'd noticed how my mom was so talkative yet my dad was extremely quiet. "Something's wrong. It's like Mom's trying to hide something."

My parents lived with us for ten weeks that winter. Although Dad interacted with our kids some and took a few short walks with me around the neighborhood, he still seemed distant and depressed much of the time. Trying to bolster his spirits, I called a friend who worked at Dodger Stadium to get game tickets. At dinnertime, I announced, "Dad, I got tickets to the baseball game tomorrow. Tommy Lasorda and Vin Scully will be there to greet us."

When Dad heard the names Lasorda and Scully, his eyes seemed to light up. I told him to be ready at noon the next day, and I would come home early from the clinic to take him to the game. When Dad was ready for bed, I reminded him again, "Now, Dad, that's tomorrow at noon."

My father smiled faintly and repeated verbatim what I'd told him, "Tomorrow at noon. I'll be ready, Ole Sox."

After seeing patients at my clinic the next morning, I hurried home to pick up my dad. To my dismay, the house was empty. I waited, and waited, thinking my wife took my parents for an early lunch and they'd be back any minute.

When he finally arrived home at five P.M., I was upset. I asked Dad if he'd remembered that I had tickets for the Dodgers game that day. Looking puzzled, my dad replied, "A baseball game?"

That's when it hit me: This was not just a miscommunication. Nothing I had said about the game or seeing his old friends, Lasorda and Scully, had registered in his mind the night before. This was more than a memory glitch; this was quite serious, even frightening.

Seeing my concern, my father tried to make light of his error, saying, "Oh, c'mon, Michael! It wasn't that important. I just did not want to put you out."

Problem was, *Michael is my brother's name—not mine.*

At that moment, my father was sitting in the same room with me. I could see him. I could touch him. But for the first time, I realized that I could no longer reach him. I walked over to Dad and asked him to close his eyes. Just as I'd done on countless patients, I rapidly tapped my index finger just below Dad's nose (above his upper lip). The response devastated me, as his lips immediately "pursed," as if to make a kiss. The pursing sign was not random! Rather, it's a clear neurological reflex that confirmed my dreaded suspicion: *Dad had Alzheimer's disease.*

I could no longer deny the obvious. After all, I am a trained neurologist, and my own father was a textbook example of the early stages of Alzheimer's. How could I have been so shortsighted, thinking that this was just depression and that he would snap out of it?

As my mind raced with ways to help my dad, I remembered an old professor of mine saying, "When you do not want to *see* the obvious truth, it will catch up to you. You will *feel* it, when it runs you over."

At that moment, I felt like a freight train had slammed my entire body into the ground.

Sometimes it's hard for people—even physicians like me, who treat patients daily—to relate to Alzheimer's disease until someone very close to them is suffering from it. As much as we might joke about Alzheimer's when we misplace the television remote or forget a loved one's birthday, the disease is devastating. Alzheimer's disease destroys the brain over a period of time, yet you never feel it happening, and there are virtually *no warning signs.* By the time there are obvious symptoms like my dad exhibited, Alzheimer's has progressed to a point where there's no turning back. My dad had not died. But he'd never again be the dad I had loved and cherished.

THE GREAT AMERICAN EPIDEMIC

I began seeing signs of an Alzheimer's epidemic about a decade ago, when younger men and women in their late fifties and sixties started coming to my neurology clinic with vague complaints of memory loss and confusion. Usually, after a few neurological exams, I sadly diagnosed them with the early stages of Alzheimer's disease. These adults were some of the movers and shakers of Los Angeles: CEOs of Fortune 500 companies, Silicon Valley executives, trial attorneys, police officers, and physicians—even doctors I'd consulted with for decades. As typically happens with Alzheimer's, in less than three years after the initial diagnosis, I'd face these same men and women in local nursing homes, many curled up in fetal positions, crying out and pleading for attention as I made my daily rounds. While their memories of the past were depleted, most were fully aware that they were alone and afraid, abandoned by family members and friends as they waited to die.

As a neurologist, I've always been passionate about finding a treatment for Alzheimer's disease. But my professional stance on treating Alzheimer's took a very personal turn when my father was diagnosed with this brain disease. As I helplessly watched my father go through the three horrifying stages of Alzheimer's, I desperately searched for solutions.

In my quest, I sorted through the mounting stacks of scientific literature, trying desperately to identify the real risks—and solutions—for people to prevent Alzheimer's disease, and I came upon a conundrum. Naturally, I had expected physicians to have a lower incidence of Alzheimer's. After all, we doctors understand lifestyle habits and behaviors that increase our risk of chronic and terminal diseases, and we should be the most disciplined in maintaining a healthy lifestyle. Yet after diagnosing a close friend and physician-mentor with Alzheimer's disease at age sixty-two, I realized that over a four-week period, I had also diagnosed two other physicians in their early seventies with Alzheimer's. That's when I began looking for commonality among patients in my practice who had Alzheimer's disease. As I made lists of patients and their former occupations, I found that doctors, attorneys, and police officers all appeared to share a similar fate—early cognitive decline or Alzheimer's. But what did these professional men and women have in common?

Then one day, a devoted wife of an Alzheimer's patient, a former night watchman, tearfully related that their home was in foreclosure. The reason? Their medication bill for anti-Alzheimer's drugs was almost $300 a month while they received only $900 from Social Security. Like so many other spouses of Alzheimer's patients, she would do anything in the world to bring her spouse back to normal. In the attempt to give her husband the latest "gold standard" drugs for Alzheimer's symptoms, she maxed out two credit cards to buy the medications. The angst I felt at that moment was overwhelming. She'd gone deep into debt to pay for highly touted and expensive medications that did absolutely nothing to stop the progression of Alzheimer's disease as promised. Further review of journal literature revealed several large British studies that confirmed my suspicions: *The anti-Alzheimer's drugs simply did not work.* At that moment, I had a dilemma, as I realized that I had been treating the wife and myself—but not the Alzheimer's patient. In doing so, I had brought the elderly couple to the verge of bankruptcy.

Around the same time that I recognized there was no pill that could revitalize a patient's memory and cognitive function, I met with some brilliant colleagues who were involved in stem cell research. These highly educated men and women made outlandish statements that former president Ronald Reagan could have been saved from the grips of Alzheimer's disease, if only he had been treated with stem cells. I had studied all the literature on stem cells. I attended the same lectures and symposiums on stem cell research as these colleagues did. I saw nothing to indicate that stem cells could reverse Alzheimer's disease. I knew their promises about stem cells reversing Alzheimer's were a complete ruse, and I was angry. As these doctors and scientists gave false hope to the former president, to his family, and to our nation, they were deceiving the very people who counted on them the most, perhaps for the same reason I had prescribed medications for Alzheimer's patients that gave only temporary relief at best. It became ultimately clear to me that there was *no magic bullet* on the horizon; no good treatment for Alzheimer's symptoms; no cure for Alzheimer's disease. That's when I became determined to find a solution: *an Alzheimer's prevention . . . not a cure.*

A meticulous analysis of the growing volumes of scientific research on the brain and diseases clearly revealed that Alzheimer's could be prevented in 70 percent of patients and delayed up to ten to fifteen years or more in those genetically predisposed. However, in order to prevent or delay Alzheimer's, I realized, there were three major problems I had to solve:

1. How to eliminate the destructive free radicals that destroy the brain;
2. How to balance the hormones and neurotransmitters that become imbalanced with aging; and
3. How to develop a diet and lifestyle program that individuals would use daily in their quest to prevent Alzheimer's.

I found convincing evidence that pointed to specific nutrients, enzymes, and combinations and proportions of foods that could help battle the free radicals and also assist in balancing hormones and neurotransmitters. A deluge of brain studies convincingly showed that physical exercise and neurobics (brain exercises) could offset the natural decline in cognition by increasing the brain's reserves. We could actually have bigger brains and improved memories with aging! All of this confirmed that while there was no Alzheimer's cure, our diet and lifestyle habits play a preeminent role in prevention. Elated by this revolutionary yet simplistic discovery, I was still haunted by one major concern: What could help people to stay compliant to a lifestyle program that was shown to prevent Alzheimer's, especially when men and women who had been so self-disciplined as physicians, attorneys, and police officers were not compliant—and were hit first by the Alzheimer's demon?

I will always remember that serendipitous early morning in mid-April 2001, when I realized that my Alzheimer's patients had three definite lifestyle characteristics in common:

1. *Poor-quality sleep* (they had jobs that required them to work eighteen-hour days or be on call throughout the night);
2. *Unpredictable stress* (they stayed in a state of "high alert," waiting for the next beeper from the hospital ER or courtroom or radio alert from police headquarters); and
3. *A sedentary lifestyle because of time deficiency* (devoted to work, they took no time for themselves to exercise or even relax).

At the same time, leaning on my studies in neurology and psychiatry, I realized that in order to keep our neocortex (thinking) brain in control of our limbic (emotional) brain, we had to maintain sufficient levels of dopamine and serotonin—neurotransmitters that are produced when we *get quality sleep, exercise, and meditate.* With high levels of the neurotransmitters serotonin and dopamine, our neocortex brain controls our decision making, which is necessary to follow any type of structured diet and lifestyle program. Thus, controlling stress, getting adequate sleep, and exercising daily are essential to keep our hormones and neurotransmitters balanced in the aging process. With our neocortex (thinking) brain in control, we are far more likely to be compliant to a healthy-brain lifestyle program and far more likely to prevent Alzheimer's disease.

Getting quality sleep and coping with stress in a positive manner is essential to maintaining cognitive function later in life. Given poor sleep, an abundance of the stress hormone cortisol in the body, and no time to exercise or relax to reduce stress, it became clear why my dear friends and patients—the physicians, lawyers, and police officers—were among the first of their generation to succumb to Alzheimer's. Controlled mostly by their limbic or emotional brains, these grand men and women lived in a constant state of high alert and were unable to achieve the balance and self-discipline it takes to adhere to daily habits that are vital for optimal brain health.

Right this minute, about 50 percent of nursing home patients in the United States have Alzheimer's disease. With the coming "Age Wave" (the 78 million aging baby boomers heading for retirement), I predict that if we don't sound the alarm about Alzheimer's right now and make the necessary diet and lifestyle changes, this period will go down in history as the Great American Epidemic of the Twenty-first Century.

How This Book Can Help

As you read this book, you will see that it quickly moves beyond a discussion of Alzheimer's disease and on to page after page of groundbreaking findings, real patient stories, and specific strategies you can use today to build a bigger brain, improve your memory, and prevent Alzheimer's disease.

Here's how I'll accomplish this:

In Part I: Understanding Alzheimer's Disease, I will give you the full description of Alzheimer's and cognitive decline—from the signs and symptoms to problems associated with dementia and how the patient and family are all impacted negatively. Starting with Chapter 1: Is Alzheimer's in Your Future?, I will continue to share a very personal and tragic story about my father's diagnosis with Alzheimer's and how his illness was the key impetus in my search for ways to prevent Alzheimer's disease. I'll introduce you to the three stages of Alzheimer's and the specific limitations patients have in each stage. I believe you must fully understand Alzheimer's—how it manifests in the brain to wipe away all memory and thoughts—in order to feel motivated to follow the 4-Step Anti-Alzheimer's Prescription. In addition, I will discuss mild cognitive impairment (MCI), a symptom complex that is a harbinger of Alzheimer's disease. Yes, you can prevent this brain disease from happening to you, if you start the program before Alzheimer's has begun its dirty work in silence—years before symptoms are evident.

In Chapter 2: Alzheimer's Disease: Are You at Risk?, I will teach you about the link between genetics and Alzheimer's disease, and then we'll focus on other risk factors, including some health conditions that you might have right now. I'll explain some simple ways to get in control of these risk factors so you can live longer and smarter. After you review the risks that increase your chances of getting Alzheimer's, you'll take the Alzheimer's Risk Profile in Chapter 3: Test Your Real Brain Age. This 25-question quiz focuses on your current health and habits. After scoring your quiz, you'll have a good idea of your Real Brain Age, a predictor of your risk for Alzheimer's disease.

You'll start the 4-Step Anti-Alzheimer's Prescription in part II, chapters 4–7. In Step 1: The Anti-Alzheimer's Diet, you'll learn about the latest findings on foods, nutrients, and brain health. I will elaborate on the Golden Dozen, specific foods I've identified that fight free radicals in the brain, balance hormones and neurotransmitters, and actually boost brain health. I will explain a method of eating each meal using the Harmonic Method (⅓ proportions of good fats, complex carbs, and lean proteins at each meal). By following the 28-day menu and recipes (in the appendix), you will be taking active measures to reduce your risk of Alzheimer's and other chronic illnesses, and increase the health of your brain.

Aerobic and anaerobic exercises are the focus of discussion in Step 2: Brawn Boosters. I'll give you a very simple plan to incorporate "the 3 *Ss*" in your daily lifestyle with *stepping, strengthening, and stretching.* I believe most baby boomers are too busy to work out at fitness centers, so my program can be done anywhere, at any time—and it's vital to balancing hormones and keeping your brain healthy and your body fit.

In Step 3: Brain Boosters, you'll learn why we must find ways to challenge our mental capacity every day. Using the many neurobics I provide, you'll be well on your way to building a bigger brain reserve and improving your memory. Whether you choose to focus on memorization, computation, analyzing poems or prose, or playing Nintendo's Big Brain Academy, I want you to plan time each day to keep your brain alert and active.

While most people accept stress and lack of sleep as a natural part of living, I believe that these two sentinel risk factors for Alzheimer's disease must be controlled before you can attempt to change the other risk factors. In Step 4: Rest and Recovery, I explain the revealing studies that link an excess of the stress hormone cortisol and lack of sleep with a greater risk of obesity, cardiovascular disease, diabetes, and Alzheimer's disease. In this last step, I'll help you to find your "circle of quiet" as you focus on rest and relaxation so you can calm the mind and help it stay strong and healthy for years to come.

In Part III: Diagnosis, Treatment, and the Future of Alzheimer's Disease, I will explain all about diagnosis and treatment of Alzheimer's. In Chapter 8: Is It Alzheimer's?, I'll help you gain insight into how an Alzheimer's specialist rules out other disorders as he or she narrows the diagnosis of Alzheimer's. I'll also give you some important questions to consider before you select a doctor to diagnose and treat a brain disease for yourself or a loved one.

Treatment varies with Alzheimer's disease, and I'll explain the most commonly prescribed medications in Chapter 9: The Latest Medical Therapies for Alzheimer's Disease. I'll explain medications for all stages of Alzheimer's, as well as specific antidepressants, antipsychotics, anti-anxiety medications, and medications for insomnia or other sleep problems, along with the effects and side effects in someone with Alzheimer's. Should you or a loved one need these medications, it helps to have insight into how the drugs work.

I want you to move quickly from understanding Alzheimer's disease and the specific risk factors in the first three chapters to the very important four-step program in part II. By starting right now to learn more about Alzheimer's disease and your personal risk for Alzheimer's, you can protect your brain's health and the health of your loved ones. To that end, this first part of the book gives you the necessary groundwork to understand Alzheimer's disease, and it sets the stage for the four-step prevention program and an understanding of how the disease is medically treated.

What Is Alzheimer's Disease?

Dementia is an illness characterized by loss of memory. While most elderly adults (60 to 80 percent) with chronic dementia have Alzheimer's disease, there are other common types of dementia such as vascular dementia, dementia with Lewy bodies, Parkinson's disease with dementia, alcoholic dementia, and even reversible dementias from misdiagnosed depression or side effects of medication.

Alzheimer's disease, a subtype of dementia, is the most common type of dementia. With Alzheimer's disease, there are physiological changes in the entire brain with amyloid plaques, a buildup of proteins that cause loss of neurons. There is vascular damage that destroys the very cells that store our memories. The Alzheimer's brain also has fibrillary tangles, insoluble clumps of twisted fibers that build up inside the neurons and between the dendrites (the communication wires) and cause memory, learning, and creativity to cease (See figures 1.1 and 1.2).

Alzheimer's was first recognized in 1906 when a German neurologist, Dr. Alois Alzheimer, examined the brain of a woman who had died after years of progressive dementia and described an autopsy with nerves caked with sticky plaque and filled with tangled fibers, which are still the characteristic findings in patients today.

Like most degenerative illnesses, Alzheimer's disease is defined by its stages that affect *cognition, behavior, and activities of daily living,* including:

1. Changes that affect cognition—Executive function (planning, sequencing, orchestrating, anticipating, and accepting delayed rewards), Memory, Attention, Intellect, and Language (changes that I call EMAIL, their acronym);
2. Changes that affect behavior, and psychological changes such as indifference, anger, illusions, hallucinations, and paranoia; and
3. Changes that affect activities of daily living such as errant driving, leaving the gas on the stove "on" and unattended, bowel

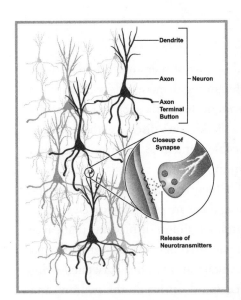 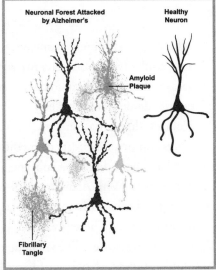

Figure 1.1 The Neuronal Forest and Figure 1.2 Alzheimer's and the Neuronal Forest—The neuronal forest contains neurons, axons, and dendrites and their synapses. The brain has over 100 billion neurons, 900 billion supportive glial cells, over 100 trillion branches, and up to 1 quadrillion (1,000 trillion) receptor sites. This forest contains the signals that form the basis of our thoughts, memories, and feelings. Nerve cells connect at the synapse. An electrical charge stimulates the synapse to release a chemical called a neurotransmitter (e.g., dopamine, serotonin, adrenaline, and acetylcholine).

and bladder incontinence, and imbalance and falls due to a loss of automatic memory to walk (apraxia).

Mild Cognitive Impairment (MCI)

Aging and dementia are in particular focus today. But when is it just aging and when is it dementia? When is it more than just absentmindedness, yet not Alzheimer's? Is there a premonitory state, a category of people who are in particular danger of developing Alzheimer's disease? The answer is yes. Though not all people exhibiting symptoms will develop Alzheimer's, the majority will. Concentrating our efforts on this group may be important for prevention.

Dementia comes from two Latin words: *de,* meaning "out of"; and *mentis,* meaning "mind"—or "out of one's mind." The practical difference between natural aging and dementia is that an aging mind still has the ability to live independently. The demented mind, even with its best effort, cannot. The aging mind recognizes reality and the demented mind increasingly loses its grasp on reality.

Recent research has identified a transitional state between normal cognitive decline and Alzheimer's disease. We call this transitional state mild cognitive impairment (MCI). Studies have estimated that those fulfilling the criteria for mild cognitive impairment develop Alzheimer's at a rate of 17 to 20 percent per year. However, not all with mild cognitive impairment develop Alzheimer's. Recognizing this group is important for both medical treatment and practical reasons. The practical reasons are that these individuals can make the proper choices for their future health care and life before they lose their ability to do so. Medical treatment, risk mitigation, and prevention are all crucial. Family, social, and financial support can help them to optimize their lives, and delay early entry into a nursing facility.

Reverend George, well-known and loved at the hospital, told me the story of how he first met his wife in Brooklyn back in the days when horse-drawn buggies still carried groceries to people, before the huge supermarkets that we have today. As George described the sounds of wagon wheels, the sights of the red clay buildings, and the smells of fresh fruits that permeated the air of Brooklyn, it seemed as if it existed right at that moment in the hospital cafeteria. I began to feel increasing anxiety as George recounted these wondrous events repeatedly—at least ten times over a month-long period.

Reverend George still drove himself to the hospital, visited his patients diligently, and was no danger to himself or to others. However, everyone had started to notice these recurring stories, along with his name- and word-finding difficulties, and a decrease in detail and personal hygiene. In addition, George relied heavily on his daily planner to remember his schedule. George's great wealth of history and his enduring pastoral empathy certainly overshadowed any shortcomings, except for those who knew him as intimately as I did.

Since I had moved from New York to California, Reverend George had become like a father to me. Over the years, he cultivated my interest in bioethics through casual references to books he thought I might find interesting. Soon, our early morning visits at the doctors' dining room for breakfast became a tradition we looked forward to. One day, I was called to the emergency room after George had failed to show up for breakfast. He had been in a car accident that morning after driving through a red light. As I look back at the classic signs of mild cognitive impairment, I now realize Reverend George had started to exhibit them four years earlier.

In those who have mild cognitive impairment, their first memories of recent events become vague. They start to rely on favorite old stories, as George did, which predominate their conversation. Behaviorally, they become less fastidious, occasionally needing to be reminded about their hygiene and grooming. Although their judgment, perception, and reasoning are intact, there is a measurable difference noted by family and friends. Lastly, when comparing those with mild cognitive impairment to age-related contemporaries, their reaction time and memory performance most certainly lags.

The dividing line between mild cognitive impairment and Alzheimer's is really not a line at all, but rather a gray zone often demonstrated by an event that clearly indicates an inability to function independently or safely. This event may be as apparent as an accident or a sudden episode of disorientation or hallucinations, or as subtle as forgetting an appointment. Once this transition begins, it slowly but insidiously worsens, with infrequent peaks of lucidness ("Look! Mom's back!") and deep valleys of sudden deterioration.

Stage 1: Difficulty with Memory and Learning

The first stage of Alzheimer's is typified primarily by cognitive changes affecting short-term memory, memory retrieval, and the ability to learn. Short-term memory recalls recent events such as what you were told several minutes before or what you ate for breakfast or the specific commitments you have for that day. Memory retrieval is the ability to recall short-term memories or long-term memories such as the name of a friend or the name of the company that you worked for in the past.

In stage 1 Alzheimer's, people forget important appointments and routinely forget where they place things. Life becomes a constant "search and rescue" mission for those who live with the Alzheimer's patient. Behaviorally, patients become disinterested in normal activities and they call friends and family members less frequently. When they do make phone calls, they talk briefly. As the disease progresses, Alzheimer's patients forget they just called and will ask the same question, such as "When is the doctor's appointment?"

Toward the end of the first stage, Alzheimer's patients may have some paranoid ideation, which is a distorted perception that someone or something is trying to harm them or take advantage of them. They often become accusatory, asserting that family members and friends are stealing from them or spying on them.

Mild cognitive impairment gives hints that the brain's cognitive reserve is drying up. When the brain's reserve "runs dry," the person will experience the classic symptoms of memory loss, difficulty learning, loss of smell, and loss of interest in usual activities, and their ability to function independently is impaired. The story I shared of my father being withdrawn, forgetful, and depressed is quite typical of MCI; his confusion and forgetting an obvious event signaled Alzheimer's disease. So many times, a doctor may misdiagnose an Alzheimer's patient with depression, and prescribe antidepressants, missing a critical time to treat and prepare the patient and family for the eventual decline.

Experiments with animals have shown a precursor of mild cognitive impairment may be a loss in olfactory cells and connections—structures located on either side of the forebrain involved in processing odors.

I will never forget getting into the car with my father and being overwhelmed by the smell of rancid meat he had left in the backseat, a year before he came down with obvious signs of Alzheimer's disease. (We conveniently explained it away as a respiratory infection.) Neurologists have smelling tests available that can determine these changes.

The pathological changes in Alzheimer's first hit the hippocampal area near the temporal lobe more severely than other areas of the brain. The hippocampus is our brain's center for storing and retrieving memory and learning, but it's not the only area to do so.

Reflex Tests Indicate Alzheimer's

On a neurological examination, there are very specific signs of Alzheimer's that are often ignored by most physicians (but well known by neurologists!). To elicit these signs in patients suspected to have Alzheimer's disease, I test the Snout, the Glabella Tap, and the Palmar Mental Reflex. These tests elicit reflexes normally seen in newborn infants that help them root, suck for feeding, protect their eyes, or bring food to their mouths. As the frontal lobes develop in newborns, these primitive reflexes disappear. Then, as the brain degenerates, these reflexes reappear again. Another primitive reflex allows an infant to hold on to his mother. When this "grasp" reflex reappears, usually in severe Alzheimer's, it is often misinterpreted as a sign of violence or stubbornness, for the demented person will not let go when you take his hand or will cling to an object placed in the hand.

Tip of the Tongue Inability Is Common

There's a loss in recall or what I refer to as "tip of the tongue inability." As an example, a patient in this stage might talk around something because he cannot remember the person's name, saying, "You know, it's the guy who shares space with me," rather than, "My colleague, Dr. Smith."

To make the premonitory recognition of Alzheimer's, doctors consider a number of signs, including lack of initiative, lack of interest, difficulty managing one's checkbook, an abandonment of pleasurable pursuits (such as golf, tennis, gardening, cooking), and finally seeking solitude to hide one's deficits.

The Reservoir Slowly Dries Up

I often explain to my medical students that MCI (mild cognitive impairment) is similar to an eight-cylinder car running on five cylinders. While it runs, when you ask the car to perform fully, it sputters, halts, and finally collapses. Another analogy to understand the difference between MCI and Alzheimer's is to think of the brain as a reservoir that is slowly drying up. Each day the "faucet" turns on, water still flows, but it does so with much less force and volume. The loss of water (reserve brainpower) occurs so slowly that one barely recognizes that it's happening. This is MCI. Then one day, the reservoir hits a critical level and no more water comes forth. This is the onset of Alzheimer's. The brain has lost its reserve and can no longer make sense of the world around it. It is at this time that the individual's brain fails. It's at this time when even a well-trained neurologist can no longer deny the obvious.

Stage 2: Irrational Thinking, Poor Insight and Judgment, and Behavioral and Psychological Changes (Paranoia, Agitation, Delusions, and Hallucinations)

Dad's Dementia Worsens

Two years after my dad was diagnosed with Alzheimer's disease, I met my mother and him at LAX. Holding on to my mother's arm, my father seemed tentative as he walked from the gate. When he saw my face, he grinned and said, "Vinny Ole Sox!"

With relief, I responded, "Daddy Ole Sox!"

Then, unexpectedly, after I embraced my dad, he looked me in the eye and asked me how old I was. To test his thinking skills, I asked him to tell me the year I was born. Dad quickly said "1951," which was the year my younger brother was born—not me.

Then I asked my dad what year it was now, and he responded, "Why it's the year after the last one."

When I persisted, he became extremely irritated and asked if this was an "interrogation." My mother immediately gave me a worried look and then commented that the weather was beautiful as usual.

The next morning, I checked on my parents before going to the hospital and noticed that my mom was shaving my dad. That evening when I arrived home, I saw my dad wandering in the living room, while the other family members were watching television in the den. I walked over to him, and he grasped my hand and would not let go. "I don't know what is happening to me. I was just in the bedroom, and I got lost. Where am I?"

Dad hugged me like a child who had just found his parent after being lost in the forest for hours. He whimpered a bit, saying, "I'm so scared. Where is Rose [my mother's name]? What have they done to her?"

Truth is, I had witnessed this scene a hundred times with patients and their families, but I'd never personally "lived it" until that moment. Despite knowing what was happening to my dad's brain to cause this tragedy, I was incapable of stopping my own father's brain from disintegrating. Right then, I saw this once vital man evaporate before my eyes. I could not stop it and I felt so helpless.

The following evening, my wife mentioned that Dad had told her little people had stolen his wallet and were hiding in the garden. He also thought the year was 1950 and asked when I would be getting home from school.

Later that night, I awoke out of a deep sleep, hearing my mother pleading with my dad to come to bed. I followed the voices into the kitchen and turned on the light. My parents stood as I had never seen them before—my father agitated, pacing; my mother with her arms out, imploring.

"Dad won't go to bed because he thinks he has to help his father at the store," Mom cried. "I can't make him understand that his father died fifty years ago, and we are in California."

I put my arms around both of them and reassured my dad that I would care for the store.

After I tucked my dad in bed, he awoke an hour later, repeating the same longing to find his dad. My mother pleaded with him, but my dad was lost somewhere in his past. The separation of past, present, and future had now melded as one in Dad's mind. This is a memory—a nightmare—I can never erase. It's part of the reason why I wrote this book.

Sundowning Is Common and Frightening

As I've seen repeatedly in hundreds of patients—and also in my own father—in stage 2 Alzheimer's, not only is the short-term memory significantly impaired, but the patient experiences periods of confusion we call "sundowning." Sundowning is named so because it occurs at dusk. As everything darkens, the person's familiar surroundings disappear, causing confusion and severe anxiety. We've all experienced similar fears being in strange places in the dark. I tell families if they want to understand their loved one's fears, imagine how they would feel trapped in a basement or attic with all the lights out, or walking down an unfamiliar street without lights. Yet for the Alzheimer's patient, it only takes a dimming of the light to bring on tremendous fear and disorientation. Finally, in stage 2, as brain cells lose connections, there are no longer specific "rooms" for past, present, and future thoughts. The thoughts meld together and become as one, as the walls in the brain that normally separate the past, present, and anticipated future disintegrate.

These individuals are no longer able to distinguish family and friends from foe. As patients transition into stage 2, they often become a danger to themselves and to others—one of the more frequent reasons patients with dementia are placed in nursing homes.

At this time, patients' paranoia may turn to hallucinations, visual or auditory, whereby they become highly agitated as they imagine something that's not actually present. Patients at this stage lose their usual social inhibitions, and usually act in an atypical manner for their personalities. For instance, I've seen many refined, sophisticated women in this stage use expletives and make sexual innuendos during normal conversation. Patients may suddenly talk explicitly about sexual acts, shocking family members or friends. Of course, these changes are in no way associated with their "real personality"—they are the result of a *diseased brain*. Along with that, I have had many spouses tell me that they are elated that the patient's personality has changed dramatically. For most of their married lives, the person was impossible to live with: aggressive, accusatory, and outright bossy. Now, the stage 2 patient is docile, calm, and agreeable. Alternately, some stage 2 patients who were meek, humble, and agreeable most of their adult lives suddenly become aggressive, demanding, and hypersexual! The behavior changes have little to do with their prior personalities. This "new" Alzheimer's personality is totally determined by the parts of the brain that work—*and don't work.*

Driving Becomes Hazardous

I recall one patient, Ryan, age seventy, who was once a highly successful business executive. Sadly, Alzheimer's had robbed him of that position two years earlier, according to his two daughters who accompanied him to the appointment at my clinic. Ryan's one hold of control over his life was his beloved 1997 Mercedes. At the end of the visit, one daughter pleaded, "Dr. Fortanasce, please tell my dad that he can no longer drive. He's becoming dangerous. Last week we were called because he couldn't find his car at the mall. Yesterday, I found a dent in his front fender and he does not know how it got there." Ryan's daughters said they'd shared this frustration with their dad's primary physician, an old golfing buddy of his who felt Ryan would deteriorate faster if his independence was taken away.

I told his daughters that even in the early stages of Alzheimer's, certain thought processes in the brain are distorted, and emergency skills are impaired. Paying attention to normal safety habits, such as looking both ways at a stop sign, is frequently ignored. Alzheimer's is not just a disease of memory—it affects everything from language to intelligence to preparatory skills, such as slowing down to stop well before one reaches a red light. A person's ability to learn and adapt to new environments and emergencies is severely impaired.

I had to do what was in the best interest of Ryan and society. I told Ryan that he could no longer drive, and then I notified the Department of Motor Vehicles. Reluctantly, Ryan agreed, and his daughters seemed relieved. Yet less than one week later, I received a call from the emergency room, and Ryan was there—confused and contused—after driving his old Mercedes into a parked car while attempting to avoid a line of preschool students crossing the street. The daughters learned that it was not enough to tell their father that he could no longer drive. They had to take the keys or have the car removed. To be honest, I'm certain Ryan forgot what I told him by the time he left my office.

Unbelievably, no one was injured. I've experienced so many other tragic endings that I will not relay them, since they'd be too painful for family members who may read this book. Unfortunately, many times the persons hurt most by Alzheimer's disease are the patients' own children, whom they nurtured for so many years.

Other Signs of Regression

It's not unusual for stage 2 patients who speak English as a second language to begin speaking in their native tongue. I remember last year when a man brought his eighty-year-old mother to see me. She had been spending the holidays with him and his family when she awoke speaking fluently in Hungarian. He later explained that she had originally come to the United States from Hungary as a young girl, and her parents and extended family spoke the language for years until they learned English.

Along with the changes in demeanor and memory loss, patients start to have sleep disturbances such as sleeping all day and being wide-awake all night. Their behavior becomes unpredictable with lability of mood—from depression to anger to apathy.

Activities of daily living also decline in stage 2. Good hygiene habits are forgotten. Patients fail to bathe. They wear the same clothes each day. Bowel and bladder problems appear, and adult diapers may be necessary. Even housekeeping tasks such as washing dishes or making beds become a major difficulty. Yet, in the midst of this, their language and motor functions remain intact—giving no one but close family members or friends a clue about their diminished state of mind.

One of the most difficult behavioral symptoms is wandering, especially at night. I was called by the police once for a patient of mine who was picked up at Chicago's O'Hare International Airport at 1:00 A.M. He had his appointment card with my name and number on it in his pocket—and no other identification. Evidently, he had walked out of his son's home where he was living, and no one heard him leave. He then paid for a ticket to Chicago from Los Angeles. Later, when we spoke, his son mentioned that his dad was born and raised in Chicago, and he was trying to return home. The most difficult symptom of Alzheimer's disease is loss of recognition of one's own spouse, children, or close friends. There is no way of describing the impact of this loss on the family; it is simply devastating.

Stage 3: Progressive Loss of Daily Living Skills

The progression from stage 1 Alzheimer's disease to stage 2 takes about one to two years. Then it's about one to two years from stage 2 to stage 3, the most horrific stage. In most cases, stage 3 Alzheimer's lasts about five to six years (or longer); the average patient lives about 8.2 years after the initial diagnosis is made.

The late stages of Alzheimer's are typified by a rapid progression in disorientation and lack of recognition. At this time, the person's ability to walk, get up from a chair, or do simple chores such as getting dressed is highly impaired. In the last stages of Alzheimer's disease, the person becomes completely dependent, needing twenty-four-hour care.

In stage 3, memory for the recent past and future often become obliterated. Recognition of family and friends, and eventually the patient's own identity, is lost. Cognitive skills are progressively lost, including language (mute), intellectual, and attention and concentration. Behavioral changes become progressively worse, going from severe paranoia and aggressiveness to loss of all personality traits. The person is just a shell.

Activities of daily living progressively decline in stage 3. The patient has bowel and bladder incontinence and must be dressed and fed by a caregiver. In time, the patient will be unable to walk or even get out of bed. Without exercise, their bodies become rigid; their muscles become fibrotic and fixed with contractures that can be severely painful. Stage 3 Alzheimer's patients need constant care for both nutrition and mobility to prevent starvation and bedsores.

STAGES OF ALZHEIMER'S DISEASE AND SYMPTOMS

Stage	Symptoms
Mild (stage 1)	Memory loss Forgetfulness Apathy Poor attention Difficulty with complex tasks Depression Difficulty with words Language problems Mood swings Personality changes Diminished judgment
Moderate (stage 2)	Behavioral, personality changes Disorientation to place and time Insomnia with Sundowning Unable to learn/recall new information Long-term memory impairment Agitation, aggression, sexually inappropriate Wandering May require assistance with activities of daily living (ADLs)
Severe (stage 3)	Agnosia (inability to recognize people, objects) Apraxia (inability to perform routine activities like dressing) Aphasia (loss of language function) Anomia (difficulty naming objects) Aphonia (difficulty with speaking aloud) Aggression Agitation Incontinence Gait, motor disturbances Bedridden Placement in long-term care

DOES AGING PREDICT DEMENTIA?

I did not give you a graphic discussion of the stages of Alzheimer's to frighten you. Alzheimer's disease is very real. In fact, millions of Americans have Alzheimer's at this very moment—maybe even people you know. It is estimated that 85 percent of people over forty have a family member or close acquaintance with Alzheimer's. While 50 percent of people over eighty-five have Alzheimer's disease, the other 50 percent do not. A little-known fact is that after eighty-five years of age the incidence of Alzheimer's disease begins to level off. The reason why may be a combination of factors. Those with genetic predisposition have already expressed it. Those who have lived a lifestyle similar to the one proposed in the Anti-Alzheimer's Prescription have resisted it.

The point I want to make is that if you know what may lie ahead in your life—such as the dreadful signs and symptoms of Alzheimer's disease—then use this knowledge as motivation to *take control of your brain's health*, and use the four steps in this book to prevent Alzheimer's disease from happening to you.

While Alzheimer's disease is increasingly common, the real problem is: *As goes our mind, so goes our ability to be independent.* Most patients with Alzheimer's are taken from their homes and all that is familiar and are moved to a nursing home for full-time medical care. I often ask my Alzheimer's patients if they are aware that they've been taken from the haven of their home. Sadly, the answer for many patients is yes.

CAN DEMENTIA STRIKE AT MIDDLE AGE?

While most adults with Alzheimer's disease are first diagnosed in their seventies and eighties, I've treated patients as young as 35 and as old as 104. I remember one patient, Elizabeth, who exhibited signs of Alzheimer's disease early at age fifty-four. Her husband and teenage daughter brought her to my clinic after noticing that she had become withdrawn, depressed, and anxious over a period of several months. An active and engaging woman who once enjoyed playing competitive tennis, she had severed ties with her lifelong friends and now spent much of her daytime hours sleeping and roaming the house at night.

When I first met Elizabeth, she had difficulty remembering the names of her siblings and close friends and had lost track of the date and year. Then within months of the diagnosis of Alzheimer's disease, Elizabeth's condition declined rapidly. She was unable to understand her husband, had difficulty speaking, and lost complete control of her bowels and bladder. Seven months after the diagnosis, her husband and daughter moved her to a nursing home for twenty-four-hour care where she lived for eight more years.

Elizabeth's case is not unusual. Early onset now occurs in about *10 percent of Alzheimer's patients* (those younger than sixty-five years and as young as thirty years of age), and this number is growing beyond expectations for unknown reasons. Sadly, within two years of receiving a diagnosis, 50 percent of Alzheimer's patients are totally dependent and unable to live alone; 70 percent end up in nursing homes within three years. These patients will live at least five more years—alone and often forgotten in that nursing home. No matter what your age or family history, the message is clear: Alzheimer's disease is a very frightening reality, particularly for aging baby boomers who are nearing retirement age.

I'm often asked why we've seen such an increase in early onset Alzheimer's disease over the last fifty years, when America is more affluent and medical care is more widely accessible than ever before.

I believe the answer is simple: Our affluent society thrives on the Western diet with a stressful, sleepless, and sedentary lifestyle. We have become our own worst enemy. I believe we have left a legacy of self-destructive habits with our children who are more obese, unfit, and discontent than ever before. If we don't make some changes, our children won't change. Remember, children do not do as we say—they do as we do.

How *The Anti-Alzheimer's Prescription* Can Help You

Now I want to turn the focus directly on you. *Is Alzheimer's in your future?* Have you thought about a time when you can no longer live alone, drive, cook, or even dress yourself? Perhaps greater than the fear of losing our mind is the fear of living alone in a dark, desolate nursing home. I believe there's a better way to spend the sunset years of life!

I've given you a simple understanding of Alzheimer's disease—the 27 signs and symptoms, the stages, and how it can happen to anyone. You're now ready to discover the hows and whys of prevention. Throughout these pages, you'll meet many of my patients—adults with Alzheimer's, healthy adults who want to prevent the disease, and family members who are on the four-step program. You'll see exactly how they've changed their own risk factors for Alzheimer's disease and are living healthier today.

My utmost goal in this book is to help you recognize your personal risk factors for Alzheimer's disease and motivate you to make the necessary lifestyle changes to keep your mind, body, and spirit healthy. *Your goal* is to take charge of your destiny—changing the risk factors you can control, so that you can reverse premature brain aging and reduce the chance of getting Alzheimer's and prevent it in your children. I'm speaking to you from my heart and mind. As a family member of someone diagnosed with Alzheimer's disease, I believe that *it's never too early* to take preventive measures. As a neurologist, I also believe that *it's never too late.*

2

ALZHEIMER'S DISEASE: ARE YOU AT RISK?

Ever since my father was first diagnosed with Alzheimer's, a question continued to linger in the back of my mind: Was the Alzheimer's a genetic condition or was it the result of my father's heart problems? Knowing about the heart/brain connection, I realized that this was a good possibility. Still, I had to know more about my genetic history. After all, we are a large Italian family with many siblings and cousins, and now we all have our own children, even grandchildren. Were we, too, at high risk for early Alzheimer's disease? When my sisters would ask me, I'd always reassure them and say no. In reality, I was uncertain.

In an effort to learn more about our ancestral history, three years ago, my two sisters, Elaine and Joan, joined my wife, Gayl, and me on a journey to Italy to find out if any relatives still remained there. There had been neither contact nor knowledge of any Fortanasces in Italy for over one hundred years. My sister Joan believed our grandfather was an orphan as our name was not after a city or famous family but simply meant "born strong." The trek through Rome to Naples and the ancient Apennine Mountains brought us to a rural, medieval Roman fortified city, San Fele, the supposed birthplace on the passport of my grandfather Vincenzo, who had come to America in 1898.

After coming almost 10,000 miles and making a Herculean effort to find the records, we got our answer. In a dusty, aged volume of birth and death records were the certificates of birth of our grandfather and great-grandfather and great-grandmother, along with the records of numerous other relatives. Once we recognized the true spelling of our name, "Fortannascere," we found relatives living (we hoped) about sixty kilometers away.

Hours later, we stood face-to-face with three older men, who introduced themselves as Michele (Michael), Vincenzo (Vincent), and Marco (Mark)—the exact names of my father, my brother, my son, my uncle, and me. However, what we learned next was far more relevant: Not only were these Italian men related to us, they revealed our genetic history. An uncle named Michael had suffered the same illness as our own father: *Alzheimer's disease*. And Vincenzo's father had died just a few years before, also with Alzheimer's disease.

My sisters, Joan and Elaine, and I left hand in hand, linked by a strong genetic predisposition for Alzheimer's. We could no longer minimize our father's Alzheimer's disease as an "isolated" condition. It is a part of our genetic heritage.

Understanding Your Risk of Alzheimer's

If you played the board game Monopoly as a child, you probably learned about taking risks: *Do not land in jail. Watch out for the income and jewelry tax. Especially avoid landing on hotels on Broadway.*

There is a major difference between the risks involved with playing Monopoly and the risk of getting Alzheimer's disease. In Monopoly, it's all up to chance; with Alzheimer's disease, it's *your choice*. As you will learn in this chapter, I believe that lifestyle habits trigger Alzheimer's disease in more than 70 percent of the cases. This means avoiding Alzheimer's disease at all cost should be your ultimate goal as you follow the 4-Step Anti-Alzheimer's Prescription.

What Are Risk Factors?

Risk factors are conditions or habits that make a person more likely to develop a disease or health condition. Sometimes risk factors can increase the chances that a disease will worsen. Other risk factors may be directly responsible for the disease itself. As an example, scientists have found that people with metabolic syndrome, also known as insulin resistant syndrome, may be at higher risk for developing Alzheimer's disease. In one study, researchers found that patients with Alzheimer's disease had a larger waist circumference, higher levels of triglycerides and glucose, and lower HDL "good" cholesterol—all symptoms of metabolic syndrome. Knowing that certain health conditions like metabolic syndrome may increase your risk of Alzheimer's, it makes sense to take steps now to control or change this serious health problem. In doing so, you will dramatically reduce the chances of getting Alzheimer's.

Sure, there are some risk factors of Alzheimer's disease we cannot influence or change, particularly the *intrinsic risk factors* including gender, age, and genetics. There are other risk factors such as obesity, hypertension, and abnormal lipids that we can control. These are the *extrinsic risk factors*. Then there are two risk factors we must do all in our power to control. I call these the *sentinel risk factors*, including (1) sleeplessness and (2) uncontrolled stress. How many times have you said "I'm stressed out and exhausted?" Only by controlling these two sentinel risk factors can you begin to change the extrinsic risk factors that increase your chance of Alzheimer's disease. These sentinel risk factors are directly linked to our motivation, determination, and ability to follow through with set goals. Without conquering these two sentinel risk factors, there will always be failure in whatever goals you set out to achieve—even with the Anti-Alzheimer's Prescription.

For example, when you feel relaxed because you wisely control your commitments and you feel rested from adequate sleep, it's easier to stay compliant to a balanced diet and exercise regimen. Thus, when you are compliant to a healthy diet and exercise program, you reduce the chances of obesity, hypertension, type 2 diabetes, and other similar problems (all extrinsic risk factors for Alzheimer's disease).

You will learn some easy strategies to control your two sentinel risk factors in the four-step program. By using the four steps as a preventive game plan, you can extend the cognitive functioning of your brain and continue to be an active and productive player in life well into your seventies, eighties, and beyond.

---

3 RISK FACTOR CATEGORIES FOR ALZHEIMER'S DISEASE

1. Intrinsic (those you cannot change)
   Age
   Genetics (family history)
   Prior head injury
   Prior heart attack

2. Sentinel Risk Factors (those you must change)
   Uncontrolled stress
   Too little sleep

3. Extrinsic Risk Factors (those you can change)
   Stress
   Sleep
   Obesity
   Hypertension
   Abnormal lipids
   Diabetes
   Sedentary lifestyle
   Homocysteine
   C-reactive protein
   Metabolic syndrome (insulin resistance syndrome)
   Other lifestyle addictions (alcoholism, drugs, environmental factors, such as pesticides, aluminum, and copper)

INTRINSIC RISK FACTOR: AGE

What We Know About Age and Alzheimer's Disease

Of all the risk factors, age is a strong determinant of Alzheimer's. While the incidence of Alzheimer's is about 1 percent in those aged sixty to sixty-five, it skyrockets to 33 percent in adults seventy-five to eighty. Then, at eighty-five, the incidence of Alzheimer's boosts to 50 percent. Women are more likely to develop Alzheimer's disease than men, even when accounting for age differentials. In study after study, researchers confirm that Alzheimer's increases exponentially with age, until age eighty-five when it begins to level off. The reason why this occurs may be that those elderly adults who are genetically predisposed to Alzheimer's already have the disease. Or it might be that those elderly adults who are not genetically predisposed have lived an "anti-Alzheimer's" lifestyle, or they may have the APOL e2 gene, which is potentially protective against Alzheimer's.

Let me explain. We know that the apolipoprotein-e2 (APOL e2) gene is associated with resistance to Alzheimer's disease, while the APOL e4 allele is associated with an increased risk for Alzheimer's disease. There is a test that can show if you have the APOL e4 gene. Along with many physicians, I encourage patients to get this test. If they do have the APOL e4 gene that shows an increased risk of early Alzheimer's, they can control their risk by adopting the four steps in the Anti-Alzheimer's Prescription, changing necessary dietary and lifestyle habits to delay getting this horrific disease. Not all of my colleagues agree with this tactic—they say there is nothing we can do. I believe this attitude can be a self-fulfilling prophecy.

Theories of Aging

There are many theories of why we age. The reasons range from degradation of bodily proteins due to free radicals, cellular toxic waste buildup, genetic mutation, and autoimmune reactions (inflammation or when your body becomes allergic to itself). A number of other theories exist, such as those that view aging as a progressive response to stress. Still another theory says that aging is a result of a change in metabolic rate or gradual deterioration of the nervous system. The truth is, the cause of aging remains mysterious even to the most brilliant scientists. That said, *the risk factors associated with an increase in the likelihood of Alzheimer's are not in question at all.* We know which factors increase the risk of Alzheimer's disease, and I've outlined them in this chapter.

Telemerase and Aging

Zeroing in on the impact of stress on aging, neuroscientists find that chronic stress can lead to disorders like depression, diabetes, and cognitive impairment in aging. The reason? Scientists think that cumulative stress may lead to shortened telomeres and decreased telomerase activity. Let me explain.

Telomeres are specialized stretches of DNA that cap the ends of human chromosomes, helping to protect them from damage and degradation. Telomerase (from the Greek word *tele*, meaning "end") is an enzyme that regulates the growth at the ends of chromosomes (bundles of DNA where genetic info is stored). Scientists now believe telomeres are critical in aging. Each time a cell divides, its telomeres get a little bit shorter. Eventually, if the telomeres become too short to divide, the cell will die. Age-related diseases and premature aging syndromes are characterized by short telomeres. Some experts have suggested that preserving the telomere may protect the chromosomes, giving the cell a longer life and slowing the aging process. In other words, the telomere seems to act as a body's "biological clock" that stops cell division and activates aging. This is believed to be the reason our anabolic hormones (growth, testosterone, and estrogen) wane with age and so do our reparative abilities. Aging is the inability to repair our bodies, with sagging, wrinkled skin being the most obvious sign.

Some neuroscientists theorize that by keeping the cells alive and dividing, it may be possible to control age-related disorders ranging from some types of blindness to cardiovascular disease to autoimmune diseases like rheumatoid arthritis or multiple sclerosis (MS). These researchers have also discovered that by adding telomerase to the chromosomes of cells, the cells continue to divide and show no signs of aging or dying. One day, this work could lead to breakthrough drugs that will stop cells from dying and preserve the functioning of body parts—even the brain—that normally break down as we age.

The Anti-Alzheimer's Prescription

I know you've read the latest media headlines on aging studies, claiming you can "turn back the clock." The reality of life is that you *cannot reverse your age*. That said, I believe that we must assess all risk factors for Alzheimer's in relationship to aging because *aging affects all the risk fac-*

tors. For instance, we characterize aging by a progressive loss of cell and tissue function, so that we become less fit to reproduce and survive. As we age, there is an imbalance between the anabolic and catabolic hormones. It's now believed that this hormone imbalance is likely responsible for many of the psychiatric and medical diseases associated with aging—and even subtle changes in hormonal patterns can exert pathological effects on health over time. As an example, elevated levels of the stress hormone cortisol can reduce lean body mass, bone density, and fat distributions that result in osteoporosis and painful fractures, metabolic syndrome, type 2 diabetes, major depression—and Alzheimer's disease. Specific habits such as diet, exercise, mental activities, social ties, and relaxation can moderate chronic stress and reduce levels of cortisol. I'll talk more about the link between cortisol, aging, and Alzheimer's disease later in the book.

INTRINSIC RISK FACTOR: FAMILY HISTORY

What We Know About Family History and Alzheimer's Disease

If you have a first-degree relative, meaning a parent or sibling, with Alzheimer's disease, this means you have a 10 to 30 percent increased risk of developing Alzheimer's, too. Also of importance are first-degree aunts, uncles, and cousins. Of course, various factors can interfere with this risk. For instance, if your relative developed Alzheimer's late in life (after age eighty-five), the risk of your getting Alzheimer's is the same as someone who didn't have a relative with the disease. That said, if your relative was young when he or she developed Alzheimer's, then your risk of getting Alzheimer's is higher. If they died early because of other diseases like cancer or heart disease, you must depend on the grandparents and immediate relatives to gain an understanding of your health history.

The Anti-Alzheimer's Prescription

I encourage you to talk with your parents and your grandparents. Ask about your family mental health history. Do they remember aunts and uncles who may have had Alzheimer's disease at a young age? Perhaps they remember a sibling or cousin who suffered with memory problems or cognitive dysfunction, yet no one ever diagnosed it as Alzheimer's. Keep a record of family members who may have had Alzheimer's disease—if not for your own benefit, then for the sake of your children and grandchildren.

On a side note, while family history is related to genetics, environmental factors can also contribute greatly to Alzheimer's disease. For instance, Cameron, forty-four, came to see me to inquire about her personal risk for Alzheimer's disease. She said she didn't smoke, drank one glass of wine with dinner, and had normal blood pressure and lipids. An active woman, Cameron was at a normal weight and in excellent health at the time.

Cameron's greatest fear was her genetic risk of Alzheimer's, because her father, Bert, was diagnosed with Alzheimer's disease at age sixty-two. Cameron described her father as a heavy drinker and smoker most of his life. He ate a diet high in saturated fat and had high blood pressure and type 2 diabetes for several decades. Each of these negative lifestyle habits and health conditions increased Bert's chance of Alzheimer's disease. Because Cameron was at a normal weight, was physically active, did not smoke, drank wine moderately, and ate a healthy diet, her risk for developing Alzheimer's was not nearly as high as her father's. I challenged Cameron to go home, review her family tree, and see if other family members had Alzheimer's disease.

When Cameron called my office the following week, she reported that no other relatives had Alzheimer's disease. If her father was the only one with Alzheimer's, maybe it wasn't Bert's genetic predisposition but rather his *unhealthy lifestyle* that destroyed his brain! I believe that lifestyle plays a strong role in 70 percent of Alzheimer's patients.

Know the Alzheimer's risk factors. Then work on changing those factors you can control so that Alzheimer's *never* becomes a reality in your life.

MY GENETIC RISK FOR ALZHEIMER'S

Family history is my own major risk factor for Alzheimer's disease, with my father showing symptoms at age seventy-two. I cannot rewrite my family history. However, I do take the following steps each day to reduce my risk of Alzheimer's disease:

- Get seven to eight hours of sleep a night after treating my obstructive sleep apnea.
- Schedule and prioritize my day to prevent overcommitments and undue stress.
- Eat a diet high in fruits, vegetables, whole grains, legumes, nuts and seeds, and fish.
- Exercise at least five times a week, thirty minutes each time of moderate exercise.
- Read, study, and continue to educate my brain daily.
- Take time to renew my spirit with meditation, prayer, and worship.
- Enjoy social gatherings with family and friends on a regular basis. All of these steps are vital to keep my hormones balanced, to decrease levels of the stress hormone cortisol that increase with age, and to reduce the chances of Alzheimer's disease—and all of these steps are *within my control*.

SENTINEL RISK FACTOR: STRESS

What We Know About Stress and Alzheimer's Disease

Stress is a sentinel risk factor. You can't avoid it, but you can learn to control it!

"Stressed out" is used to describe the impact of life's stressors. When you're stressed out, you feel increased anxiety, moodiness, distractibility, and a host of other physical symptoms. Studies now show that stress doubles or quadruples your chance of Alzheimer's disease. My professional experience tells me that it's much higher because stress is responsible for an imbalance in anabolic and catabolic hormones, even when we're young.

If you want to know how stressed you are, look down at your expanding waist—the belly fat around the abdomen. Under stress, we begin to take on an apple or lollipop shape, and this shape is linked to a greater risk of heart disease, diabetes, high blood pressure, cancer, and, yes, Alzheimer's disease. The bottom line is we now know that lifelong allostatic (stress) load may accelerate changes in the brain that can lead to memory loss.

The Anti-Alzheimer's Prescription

Revealing studies in the field of *psychoneuroimmunology* now focus on the mind and body being interconnected, and in many cases, your emotional state can determine your physical well-being. In chapter 7, I will discuss the importance of step 4, of adopting coping strategies to deal with your emotional and physical stress, including not worrying by letting go, taking things one day at a time, finding purpose in life, strengthening meaningful social ties, exercising daily, and even working on your sleep quality. Studies confirm that adults who do not employ these coping skills are more vulnerable to chronic stress, higher levels of cortisol, and subsequent illness (the body only knows what the mind tells it!). I know that chronic stress leads to both physical and mental decline years before your time. There is a reason why, after a good vacation, friends will remark how well you look and conversely how bad when you are stressed and sleepless.

SENTINEL RISK FACTOR: SLEEP

What We Know About Sleep and Alzheimer's Disease

Sleep is the second sentinel risk factor that you must control. Work with your doctor to increase your sleep to seven or eight hours nightly.

With aging, sleep disturbances are frequently triggered by stress, emotional trauma, metabolic problems, muscle and joint pain, and low-grade inflammation. Poor sleep (less than seven hours) can then lead to increased daytime fatigue with resultant diminished exercise, causing worsened physical fitness and the establishment of a vicious cycle of inactivity and sleep disturbance with physical and mood-rated symptoms. As you will continue to read, many of the extrinsic risk factors for Alzheimer's disease are linked to the two sentinel risk factors: chronic stress and poor sleep.

Sleep apnea, which occurs when you regularly stop breathing for ten seconds or longer during sleep, is one of the most dangerous and undiagnosed diseases that also increases the risk of Alzheimer's disease. Obstructive sleep apnea is caused by a blockage or narrowing of the airways in your nose, mouth, or throat that occurs when the throat muscles and tongue relax during sleep. I believe that sleep apnea may be responsible for the epidemic of Alzheimer's that we see today, and the most prevalent symptom is snoring and daytime fatigue and irresistible sleep. If you or your loved one snores, talk with your doctor about ways to test for sleep apnea. If you have sleep apnea, follow your doctor's instruction for treatment.

The Anti-Alzheimer's Prescription

In chapter 7, I will elaborate on ways to increase quality sleep to balance your hormones. Quality sleep helps decrease pro-inflammatory markers in the body and also triggers human growth hormone. Human growth hormone can decrease by as much as 75 percent by the time a person is fifty-five years old and 90 percent by age seventy. Studies show that human growth hormone deficiency leads to obesity, loss of muscle mass, and a reduced capacity to exercise—and getting quality sleep may reverse these problems in older adults. Are you beginning to understand how taking care of the two sentinel risk factors—sleep and stress—can help you conquer other extrinsic risk factors?

By the way, I'm often asked why President Ronald Reagan had Alzheimer's disease. After all, President Reagan was the picture of health and obviously ate right, exercised regularly, and had no apparent genetic predisposition. However, what he did have was tremendous stress and probably associated sleeplessness as he dealt with ongoing crises in his position as the preeminent world leader. Through no fault of his own—but perhaps as a result of his position in life—he probably was at higher risk for Alzheimer's disease.

EXTRINSIC RISK FACTOR: OBESITY

What We Know About Obesity and Alzheimer's Disease

If people keep gaining weight at the current rate, fat will be the norm by the year 2015, with 75 percent of Americans overweight and 41 percent obese. Using body mass index (BMI), a key indicator of comparing weight to height, overweight individuals have a BMI of 25 to 30. Obese individuals have a BMI of 30 and above. Problem is, obesity increases the risk of Alzheimer's disease by 300 percent for women and 30 to 50 percent for men.

Obesity represents a state of excess storage of body fat and is epidemic in the United States, with an estimated 110 million adults in the United States (about 65 percent of the population) clinically overweight or obese, having a body mass index higher than 25. Researchers have clearly linked obesity with similar pro-inflammatory markers or physiological substances in the blood that indicate disease when present in abnormal amounts. These markers of inflammation are also linked with Alzheimer's.

Obesity, particularly abdominal obesity, is associated with insulin resistance that often leads to type 2 diabetes mellitus. While the process is poorly understood, we now know that adipose or fatty tissue is a major endocrine organ that produces hormones just like other organs in the body and secretes influential pro-inflammatory chemicals, messengers that contribute to low-grade inflammation.

It is thought that as fat tissue in the body increases, the blood vessels feeding this tissue are not sufficient to maintain a normal oxygen supply and there is a localized reduction of oxygen. This triggers pro-inflammatory reactions, which can produce more cell-damaging substances in the body. Obesity can cause this response to go out of control and can add to health problems such diabetes mellitus, hypertension, heart disease, and some types of cancer. High levels of dopamine and serotonin help you control your weight. Both dopamine and serotonin depend on adequate sleep.

The increase in the concentrations of pro-inflammatory markers is now suspected to play a determinant role in the development of Alzheimer's disease. In other words, the fatter you are, the higher the levels of pro-inflammatory markers, and the increased chance of other illnesses, as well as Alzheimer's, with aging. When pro-inflammatory markers activate fat cells in the obese person to produce even more pro-inflammatory markers, a vicious cycle of obesity, inflammation, and brain disease has begun.

WOMEN AND ALZHEIMER'S DISEASE

When people ask me why Alzheimer's disease is more prevalent in women, I tell them that body fat is one of the reasons. Women have a greater fat-to-muscle ratio with a subsequent increase in pro-inflammatory markers and oxidative stress that result in cell damage.

Obesity ⇨ Inflammation ⇨ Alzheimer's Disease

The cause of obesity is simple: Our input exceeds our output. In simple words, we eat more than we exercise. However, why are some people able to control their eating while others are self-abusive? Why is it that we can control our diet at certain times, yet at other times we are completely out of control? Does our lack of dietary control correlate with times of high stress and sleeplessness? You betcha! Obesity is the result of an imbalance in your hormones. With aging, there's a decrease in the neurotransmitters dopamine and serotonin (associated with both stress and sleeplessness), which is linked to an increase in appetite and addictions. Recent research has revealed that a drop in dopamine levels signals the hypothalamus to action. Within the hypothalamus are our carnal appetites; our rage area is right next to our appetite. These are areas you'd like to keep calm—not active!

Body Mass Index and Alzheimer's

In a twenty-seven-year study of more than 10,000 volunteers, obesity was linked to a 74 percent increase of the risk of Alzheimer's; overweight participants had a 35 percent increased risk. Other findings link a higher body mass index with a decline in cognitive function in healthy individuals aged thirty-two to sixty-two years. (See page 309 to determine your BMI.)

The problem is, in most obese patients, not only do they suffer from obesity, but they also have coexisting health problems. One patient, Ginah, age fifty-one, came to see me about her own family history of Alzheimer's disease, since I had diagnosed her mother two years before. Not only did Ginah have the risk of family history, but she had her own health risks, too. Ginah was considered obese with a BMI of 31. The lab tests taken at her last doctor's exam revealed that she had high triglycerides, high LDL ("bad") cholesterol, and low HDL ("good") cholesterol. (You want your triglycerides and LDL cholesterol to be low; your HDL cholesterol should be high.) Ginah's lab results also indicated that she had metabolic syndrome or insulin resistance, which occurs when the body steadily becomes less responsive to the actions of insulin. Despite high levels of insulin, the blood sugar levels rise and eventually type 2 diabetes results. People like Ginah, who are insulin resistant, tend to gain belly fat with increasing waistlines (the apple shape as opposed to the pear shape). (See page 311 to determine your waist-to-hip ratio.)

Diet plays a role in obesity and in Alzheimer's. For example, a diet high in saturated and trans fats appears to be associated with an increased risk of cognitive dysfunction and developing Alzheimer's. Because oxidative stress is important in the pathogenesis of Alzheimer's, a diet high in antioxidants, particularly foods containing vitamin E, may decrease the risk of getting this brain disease.

The Anti-Alzheimer's Prescription

Obesity in women causes a threefold increase in Alzheimer's disease. In men, it is less than this (30 to 50 percent), but it's still significant. In fact, it's pretty much set in stone that obesity at midlife is a risk factor for Alzheimer's in later life. Here's what you can do starting today:

- If your BMI (*see chart in the appendix on page 310*) is above 30, check here _____. You are at higher risk for Alzheimer's disease.

- If your waist/hip ratio (*see page 305*) is higher than .80 for women or higher than .95 for men, check here _____. A waist/hip ratio in this range puts you at higher risk for Alzheimer's disease.

Following the specific strategies in steps 1 and 2 (chapters 4 and 5), you can change the way you eat and move around and start losing the excess pounds. In doing so, you will also reduce your risk of Alzheimer's later in life.

EXTRINSIC RISK FACTOR: HYPERTENSION

What We Know About Hypertension and Alzheimer's Disease

Hypertension, or high blood pressure, is another extrinsic risk factor for Alzheimer's. Not only can elevated levels of blood pressure cause damage to the blood vessels throughout the entire body, but also hypertension leads to heart attacks. We're now learning that problems with the heart and brain share common triggers and characteristics including inflammation, oxidative stress, and hypoxia, an oxygen deficit caused by impaired blood flow.

High blood pressure makes your heart work harder than it should, putting stress on the heart muscle and arteries and setting them up for possible injury. The effects of hypertension are profound, including the following:

- For every 7 mm Hg increase of diastolic blood pressure, the risk of stroke increases by 42 percent.
- For every 7 mm Hg increase of diastolic blood pressure, the risk of heart attack goes up by 27 percent.
- Findings show that higher levels of baseline systolic, diastolic, and mean blood pressure in both younger and older age groups are significantly associated with decline in cognitive ability (this means that young adults are just as vulnerable as older adults to blood pressure–related dementia).
- The Rotterdam study was started in 1990 in Ommoord, a

suburb of Rotterdam, among 10,994 men and women aged fifty-five and over to investigate the prevalence and incidence of and risk factors for chronic diseases in the elderly. This study found that a high diastolic pressure five years before magnetic resonance imaging (MRI), a type of brain scan, in patients with untreated hypertension was associated with *hippocampal atrophy* (which means the loss of neurons that are needed for learning and memory).

The causes of hypertension are many. Some reasons are genetic, but the majority of hypertension cases are self-induced—such as hypertension resulting from chronic stress. With increased stress, the catabolic hormones cortisol and aldosterone soar, resulting in retention of sodium with increased blood volume and resultant higher blood pressure. Sleeplessness can also increase levels of cortisol at nighttime, keeping you in a state of constant alarm and making you crave carbohydrates (a key reason for nighttime refrigerator raids).

The Anti-Alzheimer's Prescription

Keep regular tabs on your blood pressure and talk to your doctor about the best medications, if necessary. If medications are prescribed, take them religiously. Follow up with your doctor regularly to make sure the blood pressure medications are working. Adequate treatment of blood pressure has been shown by many large clinical trials to reduce the risk of heart disease, stroke, and, possibly, Alzheimer's disease.

For some people, eating a healthier diet, exercising more, losing weight, and managing stress may help to reduce high blood pressure with or without medication. Find the answers that work best for you, and work with your doctor to keep your blood pressure in a normal range.

EXTRINSIC RISK FACTOR: ABNORMAL LIPIDS

What We Know About Abnormal Lipids and Alzheimer's Disease

With increased reports of a connection between the heart and the brain, we're learning that abnormal lipids (high LDL bad cholesterol, low HDL good cholesterol, high triglycerides) not only increases the risk of heart disease, but also increases the risk of brain disease (Alzheimer's). Simply put:

- *Elevated LDL ("Bad") Cholesterol Increases the Risk of Alzheimer's Disease.*
- *High Levels of HDL ("Good") Cholesterol Decrease the Risk of Alzheimer's Disease.*
- *High Triglycerides Increase the Risk of Alzheimer's Disease.*

The Anti-Alzheimer's Prescription

Check the chart below to determine if your cholesterol levels are normal or abnormal (LDL is high; HDL is low). Then take steps to lower your LDL ("bad") cholesterol with lifestyle modifications such as reducing intake of saturated fats and cholesterol, eating more high-fiber fruits, vegetables, and whole grains, losing weight, and increasing physical activity. All of these lifestyle measures are covered in the 4-Step Anti-Alzheimer's Prescription. Now, if you use the lifestyle measures and still have high LDL cholesterol, talk to your doctor about medications that can reduce LDL cholesterol.

ATP III CLASSIFICATION OF LDL, TOTAL, AND HDL CHOLESTEROL (mg/dL)

LDL Cholesterol

<100	Optimal
100–129	Near optimal/above optimal
130–159	Borderline high
160–189	High
>/=190	Very high

Total Cholesterol	
<200	Desirable
200–239	Borderline high
>/=240	High
HDL Cholesterol	
<40	Low
>/=60	High

EXTRINSIC RISK FACTOR: DIABETES

What We Know About Diabetes and Alzheimer's Disease

Diabetes is defined as elevated levels of sugar (glucose) in the blood. Type 2 diabetes, often called non-insulin-dependent diabetes, is the most common form of diabetes, affecting 90 to 95 percent of the 21 million people with diabetes in the United States. About 54 million Americans over age twenty suffer with pre-diabetes, which causes higher blood glucose levels than normal. Problem is, a large portion of these individuals do not even know they have diabetes or are at risk for diabetes. That's because many times pre-diabetes and type 2 diabetes have no signs or symptoms until damage is done to internal organs.

In diabetes, there is either too little insulin being produced (type 1 diabetes) or there is a resistance to the existing insulin levels in the blood (type 2 diabetes). Insulin is a hormone produced by the pancreas that facilitates the movement of sugar from the bloodstream into the cells of the body.

A simple test that measures the level of blood sugar in the body can determine if you have diabetes. You have diabetes if you have a fasting blood glucose level of 126 mg/dL or higher. You have pre-diabetes if your fasting glucose level is between 100 and 125 mg/dL. If you have pre-diabetes, talk to your doctor about steps you can take right now to prevent type 2 diabetes.

Diabetes mellitus is an independent risk factor for coronary artery disease, stroke, and Alzheimer's disease. Some large studies have found that diabetes is associated with an increased risk of cognitive decline and

Alzheimer's. Findings also indicate that diabetes is associated with a 50 to 100 percent increase in risk of Alzheimer's disease and of dementia overall, and a 100 to 150 percent increased risk of vascular dementia. Scientists have determined a relationship between hyperinsulinemia and insulin resistance in the brain, as well as a relationship between insulin and beta-amyloid metabolism.

The Anti-Alzheimer's Prescription

The American Diabetes Association encourages all people over age forty-five to have a blood glucose test every three years. Early detection for diabetes will let you get proper treatment and prevent complications before there are serious problems.

KIDS AND ALZHEIMER'S PREVENTION

Diabesity is a term coined by Dr. Fran Kaufman at Children's Hospital Los Angeles. Diabesity, the triad of a high-potency sugar diet, obesity, and the subsequent type 2 diabetes, is rapidly developing in children. Parents (and grandparents) must monitor their kids' diets to create a healthier habit environment where foods high in simple carbs (sugar, white flour) are scarce, games are minimized, and exercise is maximized. Get up and get your child or grandchild to exercise with you. They'll love it, and it shows that you love them.

CONTROLLING CARDIOVASCULAR RISK FACTORS

It's not news that smoking, hypertension, abnormal lipids, and diabetes are key risk factors for cardiovascular disease. But new findings confirm that these cardiovascular risk factors at midlife may increase the risk of Alzheimer's in old age. In one study, if participants had one of the four cardiovascular risk factors at age forty to forty-four, the associated risk of

Alzheimer's was increased to 20 to 40 percent. If study participants had two of the risk factors, they were 70 percent more likely to have Alzheimer's. If individuals had all four risk factors—*smoking, hypertension, abnormal lipids, and diabetes*—they had a 237 percent (2.37 times) greater risk of getting Alzheimer's.

Treating the risk factors for cardiovascular disease and Alzheimer's using the 4-Step Anti-Alzheimer's Prescription can decrease the risk of both debilitating illnesses and let you enjoy optimal health in your older years.

EXTRINSIC RISK FACTOR: SEDENTARY LIFESTYLE

What We Know About a Sedentary Lifestyle and Alzheimer's Disease

Lack of exercise compounds the detrimental effects of eating too much. According to the U.S. Department of Health and Human Services, adults eighteen and older need at least thirty minutes of physical activity on five or more days a week to be healthy; children and teens need sixty minutes of activity a day for their health. Some recent findings show that only 30 percent of our kids in the United States get thirty minutes of exercise at one time each week. The lack of exercise in our children is a grave problem that all adults—parents and grandparents—must take seriously in order to improve the health of our nation.

We know we need physical activity to stay well and be productive. So why do 37 percent of adults report they are not physically active? Only three in ten adults get the minimal amount of physical activity each day. Adults who engage in regular physical activity and exercise are leaner and have higher levels of good HDL cholesterol and lower blood pressure than adults who do not exercise. Physical activity can also reduce the chance of obesity, insulin resistance, and type 2 diabetes—all risk factors for Alzheimer's disease.

Regular exercise also helps to decrease our stress levels. As discussed, chronic stress triggers the catabolic hormone cortisol, which is found in high levels in those with Alzheimer's disease. In animal studies, researchers have found markedly diminished cortisol levels in stressed mice who exercise. Exercise is also important for stimulating the anabolic hormones such as testosterone, estrogen, progesterone, and human growth hormone—all hormones that decrease with aging. Another benefit of exercise is that it increases neurogenesis, the creation of new brain cells, and neuroplasticity, the maturation of those brain cells into the type the body needs. With exercise, these processes stimulate some trophic hormones such as brain-derived trophic factor (BDTF) that will increase the number of cells and dendrites in the brain for connecting arms or dendrites that make us brighter in the hippocampus, the learning center of the brain. If you were a computer, exercise would be like giving your CPU an upgrade!

Exercise also increases lean muscle mass. Lean muscle burns nine to fifteen calories per pound at rest while fat burns only two calories per pound at rest. Exercise rids the body of excess glucose, especially post-prandial glucose (after a meal), so that the glucose is not converted by insulin into fat. For this reason, exercising after a meal when blood glucose is high is beneficial. Exercise has a direct effect on insulin by suppressing it. Exercise also has a direct effect on leptin and ghrelin, helping to decrease our appetite while increasing our metabolism.

The Anti-Alzheimer's Prescription

I recommend that you engage in thirty minutes of moderate aerobic physical activity at least three times a week—and every day if you can. With the exercise program I prescribe in chapter 5, there is no excuse not to exercise daily. Exercise will reduce your waist size and BMI—both risk factors for obesity that quadruples the risk of Alzheimer's in women and increases the risk by 30 percent in men. Exercise also decreases levels of cortisol and stimulates human growth hormone production. It's vital for nerve growth and the health of dendrites and synapses, which are important in retrieving memories and maintaining neurons to store in that memory.

Extrinsic Risk Factor: Homocysteine

What We Know About Homocysteine and Alzheimer's Disease

While you may know your blood pressure reading and cholesterol numbers, other indicator numbers are also very important to your health—today and in the future. One of these is the level of homocysteine in the blood. Homocysteine is a nonessential sulfur-containing amino acid that's formed by the conversion of methionine to cysteine. When elevated, this number is a warning that inflammation is destroying the body. If your serum homocysteine level is higher than 12, you may be in trouble; higher than 14—you *are* in trouble. You do not need surgery, radiation, or chemotherapy to reduce this number—only the 4-Step Anti-Alzheimer's Prescription.

Elevated homocysteine levels are a clear indicator that your hormones are imbalanced and it's about to get much worse, for homocysteine can destroy your heart, your pancreas, and finally your brain.

Homocysteine is an independent marker of the risk for cardiovascular disease. In fact, many experts rank homocysteine as equal in importance to LDL ("bad") cholesterol. Homocysteine makes blood clot more easily than normal and increases the risk of heart attack, as well as death by heart attack.

It was in the early 1960s when scientists linked homocysteine to neuropathy, a disease or abnormality of the nervous system. Since then, we've learned that serum homocysteine, similar to a sticky acid, increases with age, and elevated levels of total homocysteine are associated with atrophic changes in the hippocampus and cortical regions in the brain, particularly in the elderly at risk of Alzheimer's disease. Numerous studies now indicate that plasma homocysteine concentrations are inversely related to cognitive function in older adults—meaning higher levels of plasma homocysteine might indicate a decrease in cognitive function. As an illustration, some findings conclude that the chance of getting Alzheimer's disease is 4.5 times greater for those with higher levels of homocysteine compared to those with low levels.

The Anti-Alzheimer's Prescription

Aware for some time that high levels of homocysteine may also be associated with poor cognitive function, scientists have theorized that reducing homocysteine may also reduce the chance of Alzheimer's, or, better still, increase cognitive function. What if preventing or delaying Alzheimer's were as easy as eating the right foods? Who among us wouldn't eagerly agree to this daily regimen?

Current findings show that taking supplements of the B vitamins, including folic acid and vitamins B_6 and B_{12}, may be necessary to lower elevated homocysteine levels. I believe that homocysteine can also be managed with the right diet, including these nutrients found in whole foods:

- *Folate,* which is abundant in green leafy vegetables, oranges, and fortified cereal.
- *Vitamins B_6 and B_{12},* found in fortified cereals, potatoes with skin, and bananas.

If you don't know your homocysteine level, call your doctor and find out soon.

EXTRINSIC RISK FACTOR: C-REACTIVE PROTEIN

What We Know About C-Reactive Protein and Alzheimer's Disease

C-reactive protein is a special type of protein produced by the liver that is only present during episodes of acute inflammation. High levels of C-reactive protein are highly correlated with obesity, cardiovascular disease, diabetes, and cancer. Some groundbreaking findings now indicate that there are elevated levels of C-reactive protein in patients with Alzheimer's disease.

People who smoke, have high blood pressure, are overweight, and don't exercise often have elevated levels of C-reactive protein, whereas lean, active people usually have lower C-reactive protein levels. We now know that individuals who are prone to anger, hostility, and depressive symptoms respond to stress with increased production of the stress hormones norepinephrine and cortisol, among others. Scientific evidence suggests that an increase in stress hormones activates the inflammatory part of the immune system and triggers the expression of genes that cause low-grade inflammation, which is characterized by high levels of C-reactive protein and other pro-inflammatory markers.

In a fascinating study done at Duke University, healthy volunteers were asked to describe their psychological attributes, including anger, hostility, and depression. The volunteers did not have conditions such as cardiovascular disease or high blood pressure that would predispose them to having high C-reactive protein levels. The researchers used blood tests to measure the C-reactive protein levels and found that the volunteers who were prone to anger, had high hostility levels, and showed mild to moderate symptoms of depression had two to three times higher C-reactive protein levels than the calmer volunteers. Researchers concluded that the more pronounced their negative moods, the higher C-reactive protein levels they had.

The inflammatory response in the body destroys not only the cells intended but also the surrounding tissue. As we age, this out-of-control inflammation may be producing Alzheimer's.

Scientific methods have now revealed that both plaque and tangles seen in the brains of Alzheimer's victims contain many markers of inflammation. The plaques are full of cytokines or prostaglandins, markers of inflammation in the body. Most important of all, the microglia, the brain's roaming professional killers, are prominent in Alzheimer's type plaques. It is now thought that microglia play the key role in attacking neurons near plaques. The neurons in the vicinity get badly burned as part of the collateral damage.

The problem with inflammation and destruction in the brain is that we can't see it or feel it; the brain has no pain-sensitive neurons. So unlike arthritis, from which you might feel pain that indicates inflammation, the inflammatory attack in your brain goes undetected until Alzheimer's has you in its clutches. Not every brain will suffer the same. As you've learned by now, those individuals who are most affected are those with hypertension, diabetes, head trauma, poor diet, sleeplessness, and emotional stress.

The Anti-Alzheimer's Prescription

The good news is that by controlling the other health risk factors for Alzheimer's disease—obesity, hypertension, type 2 diabetes, homocysteine, stress, sleep disorders, and more—you can decrease C-reactive protein and other markers of inflammation in the body.

EXTRINSIC RISK FACTOR: METABOLIC SYNDROME

What We Know About Metabolic Syndrome and Alzheimer's Disease

Another related problem is insulin resistance syndrome or metabolic syndrome, a combination of health problems that includes obesity, type 2 diabetes, high blood pressure, abnormal cholesterol levels, heart disease, and polycystic ovarian syndrome (PCOD), and also increases the risk of Alzheimer's disease. It's estimated that from 25 to 33 percent of baby boomers have insulin resistance syndrome, perhaps resulting from the low-fat, high-carb diet phase of the 1970s (see page 80).

The American Heart Association and the National Heart, Lung, and Blood Institute recommend that metabolic syndrome be identified as the presence of three or more of these components:

- Elevated waist circumference
  Men—equal to or greater than 40"
  Women—equal to or greater than 35"
- Elevated triglycerides
  Equal to or greater than 150 mg/dL
- Reduced HDL ("good") cholesterol
  Men—less than 40 mg/dL
  Women—less than 50 mg/dL
- Elevated blood pressure
  Equal to or greater than 130/85 mm Hg

- Elevated fasting glucose
  Equal to or greater than 100 mg/dL

Not only does metabolic syndrome greatly increase one's risk for heart attacks, strokes, and diabetes, it's a major risk factor for Alzheimer's disease.

The Anti-Alzheimer's Prescription

If you are insulin resistant with the triad of increased blood sugar, abnormal cholesterol, and obesity—all risk factors for heart disease—talk to your doctor about a glucose tolerance test. If you have type 2 diabetes, follow your doctor's instructions about diet, exercise, and other lifestyle changes you must make to self-manage this common disease. If you do not have type 2 diabetes, focus on losing weight and exercising daily to manage your blood sugars and to prevent diabetes from happening to you. For many people, weight loss alone will help prevent type 2 diabetes and keep insulin resistance syndrome from damaging the heart, brain, and other organs.

OTHER EMERGING RISK FACTORS FOR ALZHEIMER'S DISEASE

Of course, still other risk factors for Alzheimer's are emerging. For example, studies show that people who have higher education or continue to learn throughout their lives may be at lower risk of Alzheimer's disease than those who have less education. Please note that by education, researchers are not referring to the number of years of formal education. My uncle Louie had only a third-grade education but was one of the most self-educated men I knew and lived to be a bright ninety-four-year-old. Findings show that people who continue to learn—by traveling to new places, reading books, visiting art galleries and museums, and discussing their findings with family and friends, doing crossword puzzles daily, playing a musical instrument or learning a new language—boost the size of their brain reserve and reduce their chance of Alzheimer's disease. You don't need a college education to do this!

Findings indicate that people who have a strong social network with friends and family members may have a lower risk of Alzheimer's disease than those who are loners and spend a lot of time cocooning without interaction with others. As our family unit begins to disintegrate, the elderly become increasingly isolated. This is a tremendous social issue and one that we need to address individually, as families, and as a nation.

Still other studies link cigarette smoking with Alzheimer's disease. Smoking does increase blood pressure, causing the arteries to constrict, and increases the likelihood of clots to form and obstruct the arteries. Smokers also have higher levels of LDL ("bad") cholesterol and lower levels of HDL ("good") cholesterol—all risk factors of Alzheimer's disease. In addition, researchers are trying to determine the causal role for aluminum or other transition metals (copper, zinc, iron) in Alzheimer's disease. While the link between Alzheimer's and chemicals and toxins in our food and environment has not been confirmed, there is some increasing evidence that elevated levels of these particular metals in the brain may play a role in the development of Alzheimer's. In addition, we know that Alzheimer's patients invariably have high levels of copper in the brain, which has led researchers to link copper to this form of dementia. I always tell my patients to check their vitamin labels to see if there is added copper. If so, *toss the vitamins and find a brand without copper.*

While researchers continue to seek the perfect pill to stop Alzheimer's disease, I believe there is a better way. Years of clinical research and treating patients have shown that the key intervention to stopping or reversing Alzheimer's disease is a multifaceted approach. Learning how to control your risk factors for Alzheimer's disease is essential to prevention. With some daily discipline, you enjoy your later years and live independently as long as possible.

In the next chapter, you'll find out your Real Brain Age by assessing your personal risk factors for Alzheimer's disease. If the thought of taking a test or assessment increases your anxiety, please realize that this is not a "real" test. The assessment gives you a good idea of your risk for

Alzheimer's at this moment in time. This is not to say that you can't greatly reduce this risk following the four-step program (or greatly increase your risk if you ignore some health problems and allow them to continue).

Don't be alarmed if your Real Brain Age is older than your chronological age. Once you determine problem areas in your diet, lifestyle, and health habits, you can make some dramatic changes using the 4-Step Anti-Alzheimer's Prescription and begin to plan for an active retirement with a bigger brain and increased memory.

TEST YOUR REAL BRAIN AGE

"So . . . when I will get Alzheimer's?" Susie's anxiety was all too apparent when I explained that her chance of getting Alzheimer's disease was higher than others because of her strong family history. I had diagnosed Susie's uncle (her father's brother) with Alzheimer's disease in 2006 when he was sixty. I then diagnosed her father with Alzheimer's disease in 2007 at age sixty-four. At forty-one, Susie knew how devastating the disease could be and dreaded each visit to the Alzheimer's unit at a nearby nursing home where both her father and uncle now lived.

I have scores of patients—and relatives of these patients—who call me frequently and ask "when" they will get Alzheimer's disease. Some want to get their affairs in order. Others want to have one last hurrah—a Baltic cruise or an African safari—before Alzheimer's robs them of the ability to be independent. You see, once this disease rears its ugly head in a family, a type of Alzheimer's "fear factor" immediately pervades its members—no matter what their age or health situation. I've witnessed highly successful men and women—doctors, police officers, attorneys, teachers, and corporate executives—who never once doubted their mental capabilities suddenly become neurotic, wondering "if it's Alzheimer's" when they misplace their keys, forget an appointment, neglect to keep up with their bills, or call a friend by the wrong name. Take it from me, watching someone you love undergo this unforgiving "brain meltdown" right before your eyes definitely gets your attention—and makes anyone question their own mental prowess. Nevertheless, when patients like Susie and others ask me "when" they will get Alzheimer's disease, I immediately tell them two truths:

1. Only *God* knows the future; and
2. Only *they* can take charge of their future by first checking their brain age and then starting the 4-Step Anti-Alzheimer's Prescription to reduce—and even reverse—their chances of getting this horrific disease.

IDENTIFYING YOUR PERSONAL RISK

"It doesn't seem fair," Susie confided, "that I have so many risk factors for Alzheimer's. I worry about my children and what will happen to them."

Susie began to review the list of risk factors for Alzheimer's (page 32), but had difficulty moving past her personal risk factor: genetics. Having one parent with Alzheimer's disease increases your risk of getting the disease by 10 to 30 percent; having two parents with Alzheimer's raises the chance to 50 percent.

As I explained in chapter 2, a family history of Alzheimer's disease is an *intrinsic risk factor* that you cannot change—that I cannot change. We cannot change genetics, no matter how much we'd like to. Age is another intrinsic risk factor that we cannot change. In addition, if you've had a heart attack or brain injury, you cannot erase either of these from your medical history, and both are thought to increase the chances of getting Alzheimer's disease.

Still, we have to keep in mind the risk factors we *can* and must change: the *two sentinel risk factors* (chronic stress and sleeplessness) and the *extrinsic risk factors* that cause your brain to be much older than your chronological age.

As is the case with many adults, your Real Brain Age may be much older than your chronological age, depending on the risk factors you have for Alzheimer's disease. For instance, if you are overweight, have uncontrolled hypertension and type 2 diabetes, get little sleep, and feel stressed out most of the time, your Real Brain Age may be ten to fifteen or more years older than your chronological age. (If you are fifty, your Real Brain Age could be sixty or sixty-five.) Alternately, if you are at normal weight, eat a healthy diet, have normal cholesterol and blood pressure, and fill your free time with stimulating learning experiences, your Real Brain Age could be ten years younger than your chronological age. (If you are fifty, your Real Brain Age could be forty.) For the most part, you determine the real age of your brain by how you care for yourself—with your diet, exercise regimen, educational opportunities, and rest and relaxation (the four steps).

DENIAL IS A CONSPIRACY

Whereas Susie acknowledged her higher risk of Alzheimer's disease and was ready to learn how to reduce her risk factors, *too many men and women will not admit they are at a higher risk for Alzheimer's.* In my neurology practice, adult children who bring their parents often write me notes prior to the appointment. Many notes state, "Dr. Fortanasce, please don't use the word 'Alzheimer's.' My mom (or dad) is not ready to hear this yet." Other times the note might say, "My father does not know why he's here. I told him it was about his heart—not his memory." When Susie's uncle was first thought to have symptoms of Alzheimer's disease, she handed me a note at his evaluation, simply stating, "I am so scared." Then when her father had symptoms of Alzheimer's disease a few years later, Susie handed me another note, saying, "Please help my family."

In addition, so much confusion surrounds Alzheimer's disease that almost all of my patients and their families are shocked when they first receive this diagnosis. They want to know exactly why their brains failed them. Most patients are surprised to learn that this disease did not happen over a few weeks or months but took many years to develop—years in which there were no outward symptoms of brain disease, yet their brain was prematurely aging inside. I remember when one patient, Kenneth, seventy-one, was diagnosed with Alzheimer's disease. Anne, his wife of forty-five years, wanted to know what medication I could prescribe for Kenneth to eliminate Alzheimer's disease quickly, like taking an antibiotic for a throat infection. Please realize:

There is no magic pill to cure Alzheimer's.

Nevertheless, there are scientifically substantiated strategies to *prevent* Alzheimer's disease and to keep your brain youthful, healthy, and sharp.

8 LIFESTYLE CHANGES TO PREVENT ALZHEIMER'S DISEASE

1. *Eat more fruits and vegetables.* A population-based cohort study of 1,836 older Japanese-Americans found that consumption of fruit and vegetable juices was associated with decreased incidence of Alzheimer's over seven to nine years of follow-up.

2. *Eat berries each day.* Berries contain high levels of biologically active components, including a class of compounds called anthocyanosides, which fight memory impairment associated with free radicals and beta-amyloid plaques in the brain.

3. *Eat fish high in omega-3 fatty acids.* In the Framingham study, individuals with the top quartile levels of docosahexaenoic acid (DHA) found in fatty fish like salmon, mackerel, and tuna, measured at baseline had lower rates of Alzheimer's over nine years of follow-up.

4. *Take folic acid supplementation or eat foods high in folate.* High levels of homocysteine may be associated with poor cognitive function. Some findings indicate that reducing homocysteine with folic acid may increase cognitive function.

5. *Drink a glass of red wine or purple grape juice with your evening meal.* Components in grape skins protect brain cells from the toxic effect of oxidative stress and beta-amyloid.

6. *Follow a Mediterranean style diet.* Two studies that used dietary questionnaires to assess and quantify adherence to the diet in different populations found that patients who were most adherent to the Mediterranean style diet had a lower incidence of Alzheimer's, compared with those who did not follow this diet.

7. *Control your blood pressure.* Hypertension appears to be associated with an increased risk of both vascular dementia and Alzheimer's disease.

8. *Have strong social support.* Findings indicate that an active social life and strong network of friends may help prevent Alzheimer's in later life.

WHAT'S YOUR REAL BRAIN AGE?

Take the Alzheimer's Risk Profile

While it used to be thought that Alzheimer's occurred "randomly" in the elderly population, we now know that the damage associated with Alzheimer's disease begins decades before any symptoms appear—even while you're at the top of your game at midlife!

No single approach will solve the problem of Alzheimer's disease. What is called for is an *individualized approach* that considers all the personal risk factors of each of us. To help you determine an individualized approach for you and your loved ones I've designed the following 25-item questionnaire—Test Your Real Brain Age—that analyzes your risk of Alzheimer's disease. Once you've completed the Test Your Real Brain Age quiz, you can score the quiz using the instructions on page 63.

The best way to approach the Alzheimer's Risk Profile is to take a piece of paper and write the numbers 1 through 25, and then record your answers next to the number. At the end, you can easily compute your Real Brain Age using the key. After you've scored the Alzheimer's Risk Profile, give the test to family members, friends, and colleagues, and tally their scores. By following the proven solutions, safe strategies, and self-care recommendations that I teach you in steps 1–4 (chapters 4–7), you can develop a personalized program to dramatically reduce your risk of Alzheimer's disease. Ultimately, you will greatly improve your quality of life and reduce your risk of many chronic diseases associated with aging.

On your numbered sheet of paper, write *True* or *False* for each statement below. After you take the test, calculate your score using the scale on page 63. I will continue to address this profile throughout the book and teach you how changing your diet and lifestyle habits will decrease your chances of getting Alzheimer's disease—giving you a more youthful brain and increasing your overall health and well-being.

1. I get seven to eight hours (or more) of sleep each night.
2. I eat at least five or more servings of fruits and vegetables high in antioxidants daily.
3. I eat at least one serving of blueberries, raspberries, or blackberries daily.
4. I eat baked or broiled fish high in omega-3 fatty acids (especially eicosapentaenoic acid and docosahexaenoic acid) at least three times a week.
5. I take fish oil supplements high in omega-3 fatty acids or flaxseed supplements at least five times per week.
6. I take folic acid supplementation with my daily multivitamin.
7. I take a low dose of aspirin daily.
8. I drink red wine or grape juice at least five times a week.
9. I exercise most days of the week for at least thirty minutes each time (total of three hours or more of strenuous exercise weekly).
10. I read challenging books, do crossword puzzles or sudoku, or engage in activities that require active learning, memorization, computation, analysis, and problem solving at least five times a week.
11. My total cholesterol is less than 200.
12. My LDL ("bad") cholesterol is less than 110.
13. I have "longevity genes" in my family, with members who lived to 80 and older without memory loss.
14. I am not obese (less than twenty pounds overweight for a woman; less than thirty pounds overweight for a man).
15. I eat a Mediterranean style diet (high in fruits, vegetables, whole grains, beans, nuts, and seeds, and olive oil as the source of fat; little red meat).
16. I use olive oil and spreads with *no trans fat* instead of butter or margarine.
17. I have never smoked cigarettes.
18. I have normal blood pressure.
19. I do not have diabetes.
20. I do not have metabolic syndrome (high triglycerides, central obesity, and hypertension), also called insulin resistance syndrome.

21. I do not have a sleep disorder such as snoring or obstructive sleep apnea or untreated insomnia.
22. Daily uncontrolled stress is not a problem for me.
23. I have a strong support group and enjoy many activities with friends, colleagues, and family members.
24. I have no problems with short- or long-term memory.
25. I'm ready to prevent Alzheimer's and am willing to do whatever it takes.

Calculate Your Real Brain Age

Now please go back and count how many of the 25 true/false statements statements you marked "True." Write your score on a sheet of paper and then use the following key to determine your Real Brain Age and risk of Alzheimer's disease. Bear in mind that I'd have to meet with you individually to get all of the complex medical history necessary to give you a complete picture of your risk. Here, we're aiming for a relatively simple means of tallying a score that will be easy to understand and apply to your diet and lifestyle habits.

23–25 Congratulations! You are aging well. *Subtract 15 years* **from your chronological age for your Real Brain Age.**
You are presently healthy with a youthful, productive mind. Following the four steps in this book will make your body, mind, and spirit even healthier. Unless things change in your life, your risk of Alzheimer's disease is extremely low.

20–22 Not bad! *Subtract 10 years* **from your chronological age for your Real Brain Age.**
You are doing a lot to take care of your physical and mental health. Check the specific questions that you marked "False"—and be sure to pay attention to changes you need to make. Following the four steps in this book will make your body, mind, and spirit even healthier.

15–19 OK. Your Real Brain Age is the *same as your chronological age.*
That said, you have a mild risk of Alzheimer's disease, so pay attention.

Carefully review the four steps to see what changes you need to make with your diet, exercise, mental stimulation, or rest and relaxation.

12–14 You have a moderate risk of Alzheimer's Disease. *Add 5 years* **to your chronological age for your Real Brain Age.**

While there's not a lot of disparity between your Real Brain Age and your chronological age, I'd still like you to reread the information in chapter 2 so you really understand the risks you have that increase the chances of Alzheimer's. It's important that you review the quiz and circle any of the statements that indicate some work is needed. Talk to your doctor about your Alzheimer's risk factors you have to see if treatment is indicated.

0–11 You have a high risk of Alzheimer's disease. *Add 10 years* **to your chronological age for your Real Brain Age.**

Right now, call your doctor and talk openly about health problems you have. Ask if you're doing all you can to manage these problems. In addition, read this book and flag those pages that may help to decrease your risk of Alzheimer's disease.

Balancing Your Hormonal Symphony

Throughout this book, I will refer to your "hormonal symphony" from time to time, as a way to help you understand the physiologic and hormonal changes that occur with aging and how the four-step program balances these hormones to prevent Alzheimer's disease.

First, I need you to visualize a symphony orchestra (see figure 3.1). If you've ever heard a symphony orchestra, then you've heard the combination of many instruments making completely different sounds yet coming together in perfect balance to produce a melody that's sensually pleasing. When you watch a symphony orchestra play, the string instruments—*the violins and cellos*—are to the left of the stage; to the right are the *brass and drums*. Using smooth, graceful movements, the symphony conductor accentuates one section by raising a hand while diminishing another section by lowering a hand. As a result, the symphony is in perfect harmony and you enjoy the pleasurable experience.

Your body also has a symphony—*a hormonal symphony*—that includes your hormones, neurotransmitters, and nervous system that runs in conjunction with the hormones. In your body's hormonal symphony, let's say that insulin is the conductor. Most people understand that if insulin levels are balanced, there is harmony or health in the body, mind, and spirit.

Figure 3.1 The Hormonal Symphony

The Nervous System

The nervous system, vagal and sympathetic, is part of what we call autonomic or automatic, which simply means the nervous system is not under your conscious control. The nervous system carries messages to other neurons or body parts through the various neurotransmitters such as serotonin, dopamine, adrenaline, noradrenaline, and more. The vagal nervous system, in general, slows you down and causes a decrease in your pulse rate and blood pressure. The sympathetic nervous system revs you up and prepares you for "fight or flight."

The Anabolic and Catabolic Hormones

In your body, the anabolic hormones (growth hormones, testosterone, estrogen, and thyroid) and the neurotransmitters dopamine and serotonin, together with the vagal nervous system, represent *the strings and the piano* of your orchestral symphony. On the other side of the orchestra, representing *the drums and the brass,* you have your catabolic hormones, namely cortisol and aldosterone. These hormones are associated with the neurotransmitters noradrenaline and adrenaline, and the sympathetic nervous system.

The anabolic hormones' (*violins and cellos*) effect on the body is to increase lean body mass and muscle strength, and to decrease fat. In your brain, the anabolic hormones act as antioxidants, wiping out free radicals (part of the body's defense system), increasing the strength of the immune system, decreasing the release of cytokines (pro-inflammatory markers), and adding to endurance, positive mood, and a feeling of calmness and control via dopamine and serotonin. When the vagal nervous system is stimulated after exercise, it causes the body to go into a relaxation mode. When this occurs, your heartbeat slows, your skin feels warm and flushed, your pupils relax, and muscle tension diminishes. These are all the physical signs of the mystical Zen or the feeling of being calm, contented, and in contact with nature.

Now, when the catabolic hormones, together with adrenaline and noradrenaline, are excreted by the adrenals, it causes an increase in fat production, an increase in blood glucose, a decrease in lean muscle mass, and an increase in insulin. Insulin, the conductor of your hormonal symphony, tries desperately to control this catabolic response and must work overtime, increasing its levels as it frantically tries to rid the body of the glucose by storing it as fat. When glucose is not available through recently eaten food, cortisol breaks down protein to glucose and, in doing so, reduces your much desired lean body mass. If glucose is produced in excess, insulin (your conductor) rids the body of it by storing it as fat (LDL or "bad"cholesterol). When the emergency threat remains for a prolonged time, the adrenal glands produce the stress hormone cortisol, allowing the body to prolong this emergency response. What is your body's response? It is an increase in your heart rate, cardiac output, and pupil size. Your muscles tense and soon you have a headache, muscle aches, and even fibromyalgia with deep muscle pain, insomnia, high anxiety, and depression. In the blood, there is a surge of glucose with insulin following to control it. The conductor (insulin) rises to the occasion!

There are several effects on the brain when insulin goes above normal levels. To start, an enzyme produced by the brain to control insulin and reduce glucose comes into action. This is called the insulin-deleting enzyme (IDE). Problem is, the insulin-deleting enzyme is the one that also rids the brain of the toxic amyloid protein. When insulin and blood glucose levels are out of control, the insulin-degrading enzyme works overtime to remove the insulin—instead of removing the formation of amyloid, plaques, and tangles—*the very pathology of Alzheimer's disease.* As insulin soars to high levels, it overwhelms the circuits in the brain, resulting in brain freeze. In psychology, we call this process "blocking." The area of the brain most affected by brain freeze or blocking is the hippocampus, the learning and memory center of the brain. Finally, the increase in available amyloid protein releases the deleterious free radicals that attack neuronal cell membranes, the mitochondria (the energy engine), and the nucleus itself. This "brain" damage then provokes the brain's own defense cells, the macroglia, to attack the damaged cells, thereby causing the loss of cells we see occurring in Alzheimer's disease. As you'll learn in Step 1: The Anti-Alzheimer's Diet (chapter 4), the key to preventing the buildup of amyloid protein in the brain is to keep your levels of insulin balanced—and you can do that by following the prescribed diet.

Are You Ready to Live Longer and Smarter?

About seven years ago, I became highly aware of my personal need to make some changes and take better care of myself. It was a Saturday morning. I had been on call the evening before with the usual multiple calls from the hospital and emergency room. After I threw on some shorts and a T-shirt, I stared at myself in the full-length mirror and wondered who in the world it was. Sure, it was me, Vince Fortanasce. However, it wasn't the real Vince—the one I remembered in my twenties and thirties. Ironically, several of my baby boomer friends have related similar disturbing experiences. One golf friend commented that they "sure don't make mirrors like they used to twenty years ago." Another said it took him "twice as long to look half as good" as he did at forty.

The fact is, we all age. Most of us put on some weight and add some inches (obviously not in height). Most of our bodies get misshapen from the pull of gravity (or the lure of beer, chips, and fast foods). However, that morning six years ago when I looked at the reality of my aging, out-of-shape physique, I didn't like what I saw. With great resolve to change my diet and lifestyle habits, I grabbed a popular diet book off the dusty shelf in my study and marched down to breakfast.

Immediately, I opened the book to a list of specific foods to eat and foods to avoid. My determination was unyielding, and I was in control. I vowed that I'd stick with this diet and finally take control of my out-of-control weight. My neocortex, or my thinking brain, was now calling the punches. There would be no more emotional comfort eating to please my limbic brain, no more procrastinating or thinking that I could lose weight next month. I was fully determined to do what was healthy for me. With my hormonal symphony in perfect concert, I made mental plans to take a three-mile jog after breakfast and then go by the gym later that afternoon to pump weights. I also planned to meditate and rest, realizing that if I was not stressed, I would be more compliant to the diet and lifestyle program.

As I began to prepare my oatmeal, I was in total control of my mind and body . . . until my beeper went off. It was the hospital laboratory suite. I had forgotten that I had scheduled a patient for a test that day, and my staff and other doctors were trying to reach me. Now read this section fast! Within seconds, my hormonal symphony was in disarray. Adrenaline, followed by the stress hormone cortisol, flooded my bloodstream as my limbic (emotional) brain screamed and the fight-or-flight response kicked in. I dropped the box of oatmeal and grabbed three giant doughnuts and a super-size cola as I rushed to grab my coat and dashed to the hospital. Gone was the calm reasoning of my neocortex; as my heart and blood pressure were pumping and soaring, my emotional (limbic) brain was in command. As my insulin levels peaked with each bite of the sticky glazed doughnuts, my hormonal symphony shrieked off-key with the strings and brass playing to the beat of their own drummer.

STOP! Did I know what I was doing to my body, my heart, and my brain at the time? Of course, I'm a doctor. However, just like everyone else, doctors are also influenced by their limbic, emotional brains.

Now, here's the catch: There is a concrete reason why some people succeed at being compliant to a diet and lifestyle program and the rest of us fail. This reason is not based entirely on psychological toughness, the so-called determination or "willpower" theory. Rather, those individuals who maintain compliance to a diet and lifestyle program are the ones who can go the long haul—they maintain their determination or will-power the longest and make the diet a habit, a daily part of their lives. The program is no longer a short-lived decision, like my brief commitment to diet and exercise before the emergency call from the hospital. For you to stay compliant with any change in habits, the change must become a habit—a regular part of your day-to-day routine, just like brushing your teeth, combing your hair, or other activities you do without thinking.

Psychological studies show that you must maintain a change in habits (whether diet, exercise, stress reduction, stopping cigarettes, or other) for twenty-eight days for it to become a "habit." To develop this habit, you must keep your thinking brain (neocortex) in ultimate control of your limbic, or emotional, brain. As I experienced, even the slightest disruption can cause your limbic brain to erase all your good intentions (*intentions are not habits!*). I challenge you as you read this chapter to stay focused on your goal.

WILL YOU CHOOSE TO PREVENT ALZHEIMER'S DISEASE?

It doesn't take a rocket scientist to know that success or failure depends on your ability to follow through. In fact, there is a psychological reason behind your overt decision to purchase this book to prevent Alzheimer's disease. Was this a rational decision to live healthier or to be brighter? Or was this decision made out of fear of Alzheimer's, perhaps because a loved one suffered greatly with the disease? Was it your thinking, cortical brain or your emotional, limbic brain that made this decision?

The media definitely knows the answer. They use think tanks to tell them which emotional buttons to push to motivate public opinion and response (e.g., sex sells, as does a fear of illness or death). Yet what influences the lifestyle decisions or choices we make? The answer is unequivocal: The choices we make are determined by the limbic (emotional) brain most of the time. To the contrary, willpower or the ability to stay on course and say no to our emotional brain is determined by the neocortex (thinking) brain.

The thinking brain is well developed in humans in the frontal, outer layer of the brain called the neocortex. The emotional, limbic brain is well developed in animals and lies much deeper in the brain. The neocortex is the conscious brain: the ego or superego. The limbic brain is our unconscious brain or id. Our neocortex tells us to do what's right, care for the weak, be compassionate, exercise daily, and eat berries instead of cake. The limbic brain tells us to take care of "me," avoid pain, seek pleasure, lie on the couch and watch TV, eat three doughnuts instead of fruits and vegetables. The question is how can we keep our thinking, neocortex brain in control? How can we prevent a doughnut binge when we're under emotional stress? The answer lies in our hormonal balance.

Willpower and the Hormonal Symphony

Since our brain has both rational and emotional components, we cannot dismiss either part. The control of the neocortex over the emotional brain depends on the balance, harmony, and concert between our hormones (anabolic and catabolic), neurotransmitters (serotonin and dopamine), and autonomic nervous system. It is an understanding that we cannot succeed if our rational brain does not take care of our emotional needs. If the thinking brain can stay conscious and be in control of our limbic brain's needs, we will be successful in preventing Alzheimer's and living a long, healthy, and bright life.

Problem is, as we age, our ability to maintain hormonal and neurotransmitter balance becomes increasingly difficult because of a natural loss of beneficial anabolic hormones and an increase in catabolic hormones. This is the physiological reason why our motivation wanes, our anxiety soars, and the ability of our cortical, thinking brain to stay in control diminishes. Our ability to stay in control lessens. As we age, our brain's control of sleep is disturbed and we have trouble getting enough sleep. Stress becomes an overwhelming daily issue because the brain cannot turn off its alarm (the cortisol hormone system). Then because we're stressed, we feel fatigued. To boost our energy, we binge on simple carbohydrates—chips, cookies, and sugary drinks. Yet we neglect to exercise. All of these factors are cumulative and result in hormonal imbalance or dissonance in the hormonal symphony.

When the anabolic hormones are high, the catabolic hormones are controlled, and the serotonin and dopamine levels are high, our thinking brain—the neocortical brain—will stay in control. This is when we have willpower and can comply with a diet and lifestyle program.

On the contrary, when the anabolic hormones fall and dopamine and serotonin levels fall, it's the limbic brain, the emotional brain, that prevails. When the limbic brain rules, you'll binge eat or drink, feel highly anxious or depressed, and have an inability to follow through. You'll fall prey to addictions, as you seek ways to soothe the emotional brain's needs. When the neocortex brain rules, you will stay in control, making choices that serve to keep you well.

The 4-Step Anti-Alzheimer's Prescription will help you maintain hormonal and neurotransmitter balance with insulin, the conductor of your hormonal symphony, balanced. As you do so, the thinking brain will make the decisions in your life—not your emotional brain. By making smart, conscious diet and lifestyle choices, you can slow the degradation of the brain. In effect, you can decrease your Real Brain Age (a younger brain is a more efficient, bigger brain!) and slow down your body's biological clock. I truly believe this is an exciting promise—that we can actually *choose* to prevent Alzheimer's and have a bigger, better brain. To be honest, this is *the only hope* we have right now for preventing this mind-robbing disease, so let's get started!

PART II

THE 4-STEP
ANTI-ALZHEIMER'S
PRESCRIPTION

4

STEP 1: THE ANTI-ALZHEIMER'S DIET

FOOD FOR THOUGHT

Last summer, I spent an evening at the home of former Major League baseball pitcher and two-time World Series champion manager Tommy Lasorda, where we sat in front of a huge-screen TV with the backdrop of the Los Angeles Dodgers playing the San Diego Padres. With his eyes fixated on the television screen, Tommy listened as I explained the Anti-Alzheimer's Diet to him and his wife, Jo. After about twenty minutes of my Anti-Alzheimer's diatribe, talking about the Golden Dozen (the twelve brain-boosting foods I recommend on the diet) and the Harmonic Method of eating (eating ⅓ distribution of calories from good fat, lean protein, and complex carbohydrates throughout the day), my dear friend Tommy looked me in the eyes and said, "Just do it, Vince. Give me the diet. Tell me what to eat."

If you knew Tommy Lasorda, you'd know that the only thing bigger than his appetite is his heart! Today at eighty, Tommy moves at incredible speed and is constantly on the road, whether speaking to thousands on behalf of the National Prostate Cancer Association or giving motivational talks to Little League coaches at remote baseball fields in San Bernardino County, California. Certainly not one to spend hours poring over a diet plan, Tommy wanted me to hand him a "plain and simple" version of the Anti-Alzheimer's Diet all wrapped up with a bow. "Just tell me what I have to do, Vince. Lay it out in plain language." I did that for him and, in this chapter, I'll do it for you, too.

About a month after our first discussion of the Anti-Alzheimer's Diet, when Tommy and I met for lunch, I discovered that while Tommy was eating the specific foods on the Anti-Alzheimer's Diet and in the right proportions, he had no idea why he was doing this. He didn't understand the science behind balancing insulin levels for a bigger brain or why he was eating specific foods high in antioxidants, folic acid, and omega-3 fatty acids. On the other hand, his wife, Jo, had studied all the reasons for the program and understood the scientific substantiation, including why it was important to select foods to keep insulin levels balanced for better brain health. Jo was thrilled with the Anti-Alzheimer's Diet and said she felt better than she had in years.

Did you know that a diet is unlikely to become a true lifestyle change unless you understand the rationale behind it? An intriguing study done at Yale confirmed that when the rationale of a diet plan was explained to participants, *more than 55 percent* were still compliant one month later. This number is compared to only *20 percent* of those participants who were still compliant to the diet after one month who had been given no explanation. The excitement of starting a new diet and lifestyle program may keep your interest briefly—for a day or two. Nevertheless, without long-term motivation—and an understanding of "why" you are making the changes—the plan will be lost in life's daily demands and distractions.

While we were eating lunch, Tommy reminded me of why his 1988 World Series Champions won. Was it because the team was far superior or more talented than other teams? Not at all! It was because his team used specific ingredients in the proper method—that is, Tommy Lasorda knew how to mix the lineup, recognize when a player was at the top of his game, and determine which players needed to take on different positions. Keeping his players in perfect harmony, Tommy kept the players' minds and bodies in top form, and his team won the World Series.

This same theory of "harmony" works for the Anti-Alzheimer's Diet, too. This diet plan differs from so many others because it depends on eating specific brain-boosting foods—and the right proportions of these foods, using the Harmonic Method—not just eating a low-calorie or low-fat or low-carb diet.

THE INSULIN–GLUCAGON BALANCE AND YOUR BRAIN

In the last chapter, I introduced the concept of the Hormonal Symphony with insulin as the preeminent conductor. Insulin is a hormone released by the pancreas that promotes the storage of calories, increases fat stores, and regulates blood glucose (sugar) levels in the body.

By eating the Golden Dozen and the prescribed combination of the recommended good fats, lean protein, and complex carbohydrates (the Harmonic Method), you will maintain normal levels of insulin and blood glucose levels while maximizing special hormones called glucagons. Glucagons are made in the pancreas and raise the level of glucose (sugar) in the blood by releasing stored glucose from the liver. Glucagons are essential for brain health and for keeping insulin, the conductor of your hormonal symphony, in control. Here's how it works.

When you eat lean protein (fish, lentils, skinless chicken or turkey, or soy protein), glucagon is excreted rather than insulin. The excretion of glucagons converts protein and fat into glucose. On the other hand, if you feast on simple carbohydrates (sugary pastries or candy, white potatoes, white flour, white pasta), only insulin is excreted. When glucagons use protein and fats to produce sugar, they burn many more calories than the direct use of simple carbohydrates. The consumption of refined carbohydrates and sugars is like pouring jet fuel into the bloodstream. Rather than a gradual increase in glucose caused by the proper combination of lean protein, good fats, and complex carbohydrates, we get an immediate high from refined sugars injected into the bloodstream. This kicks the pancreas to produce high levels of insulin in an attempt to regulate the high blood-glucose level. Some of the blood glucose will go for energy consumption, which is needed by the muscles and other body organs. The rest is converted to fat. So, high insulin equals increased fat.

Simple Carbs Can Lead to Alzheimer's

If you eat an imbalanced diet high in simple carbohydrates long enough, your body becomes less responsive to the action of insulin. Your blood sugar levels soar despite high levels of insulin and the excess blood glucose is converted into fat (belly fat around the waistline). This leads to insulin resistance syndrome, type 2 diabetes, elevated cholesterol, higher LDL (bad) cholesterol, lower HDL (good) cholesterol, and increased triglycerides.

High sugars or glucose in the blood will damage the nerves and small blood vessels of the eyes, kidneys, and heart and predispose you to atherosclerosis (hardening) of the large arteries that can cause heart attack and stroke, and lead to Alzheimer's disease. The high insulin levels also affect the insulin-degrading enzyme (IDE) that exists in the brain. This key enzyme removes beta-amyloid, the toxic protein that produces Alzheimer's disease. When insulin levels in the body remain high, the insulin-degrading enzyme works overtime removing insulin rather than deleting the beta-amyloid proteins. If this happens repeatedly and the insulin levels in your body remain high for a length of time, there is a slow, inexorable accumulation of beta-amyloid in your brain, which can result in Alzheimer's. This is analogous to what smoking cigarettes does to your lungs.

The Balanced Harmonic Method Protects Your Brain

The Anti-Alzheimer's Diet is based on the Harmonic Method—a balanced ⅓–⅓–⅓ distribution of calories throughout the day from good fat, lean protein, and complex carbohydrates. This balanced proportion of good fat, lean protein, and complex carbs prevents levels of insulin from getting too high. With too much insulin in the body, you feel exhausted and you get fat. Yet too little insulin in the body leaves you feeling weak and sick. When insulin, the conductor of your hormonal symphony, stays balanced in normal ranges, you feel great and live healthier, longer, and smarter! Balanced levels of insulin lead to optimal brain health.

If you've taken the health of your brain for granted up until now, I hope to convince you right now in this step to start feeding your brain the proper nutrients in the proper proportions so your brain stays vital and healthy—for the rest of your life.

WHAT IS THE ANTI-ALZHEIMER'S DIET?

The Anti-Alzheimer's Diet is a sensible diet that's high in antioxidants, omega-3 fatty acids, folate, and other crucial brain-boosting nutrients. It's also high in good fats—oily fish, olive oil, nuts, flaxseeds, and other heart-healthy fats—and lean protein and complex carbohydrates, which are distributed throughout the day to maintain normal blood sugar levels. The Anti-Alzheimer's Diet does not recommend starchy foods such as white bread, white pasta, or white rice—all foods that are digested into sugar within minutes of eating. The refined starches and concentrated sugars raise blood sugar quickly, which can lead to overeating and weight gain.

Not only will the Anti-Alzheimer's Diet help you prevent Alzheimer's disease, but it will help you manage your weight and prevent obesity, lower your blood pressure and lipids, and reduce the likelihood of diseases like cardiovascular disease and diabetes that lead to Alzheimer's disease.

The Anti-Alzheimer's Diet Is a Mediterranean Style Diet

The Anti-Alzheimer's Diet is a Mediterranean style diet in that it encourages fresh fruits and vegetables, fish and other lean protein, nuts and seeds, legumes, unrefined whole grains, olive oil and other good fats, and low-fat dairy products such as yogurt and natural cheese. With this Mediterranean style diet, you avoid red meat, sugary desserts, and processed foods.

Some background: When researchers at Columbia University analyzed the diets of 1,984 people averaging 76.3 years of age and scored them from 1 to 9, they realized that the higher the score, the closer the participant's diet was to the Mediterranean diet. The Columbia study concluded that the risk of Alzheimer's was lowest among those people with the highest scores—those diets that were most like the Mediterranean diet. For each point on the diet scorecard, the risk of Alzheimer's dropped by 19 to 24 percent—so, for example, if you scored a 6, your risk would be 19 to 24 percent lower than someone who scored a 5.

Those who scored in the top one-third had a 68 percent lower risk of Alzheimer's than those in the lower one-third.

The Anti-Alzheimer's Diet Is a "Good Fat" Diet

In the late 1960s, the National Institutes of Health (NIH) issued a report that coronary artery disease was associated with cholesterol, and cholesterol was associated with foods high in fat. Their conclusion? All fat is bad. In later reports, they warned that we must reduce or eliminate fats and even protein, which was often laced with fats, and we must maximize carbohydrates. Eat those carbs, they thought, and don't touch fat! And with little knowledge of human physiology, the NIH gave this "low-fat diet" mandate to the U.S. Department of Agriculture (USDA).

The resulting industrial machine pumped out tons of low-fat, high-sugar, carbohydrate-laden cereals, which were marketed and fed to the youth, namely us—the baby boomers. The consequence of low-fat dieting has been an unprecedented increase in hypertension, metabolic syndrome (insulin resistance syndrome), type 2 diabetes, and Alzheimer's disease in those under age sixty-five. The USDA did not consider that our bodies assimilate food no differently today from the way we did ten thousand years ago, when all we had to eat were fat, protein, vegetables, and berries—not refined sugar and white breads. Sugar and white flour stream glucose throughout the body, peaking insulin, and provoking amyloid deposits and free radicals in the brain. The excess carbohydrates and fatty acids are part of the free radicals that are like scavenger molecules attacking cell membranes and causing early aging of organs like the skin, heart, and liver. The USDA, in a rushed effort to help, created a Titanic food pyramid that has sunk many into a life of type 2 diabetes and Alzheimer's disease. We are only now seeing the tip of the iceberg.

In the Anti-Alzheimer's Diet, you'll eat monounsaturated fat that comes from oils that are liquid at room temperature, or better still from plant foods (olives, flaxseed, nuts, and avocados); omega-3 fats, the highly polyunsaturated fats found in fatty fish (anchovy, sardines, shad, mackerel, tuna, and salmon); and polyunsaturated fat that comes from oils that are liquid or soft at room temperature, including corn, soybean, sunflower, safflower, and sesame oils.

The Anti-Alzheimer's Diet Recommends Foods Low on the Glycemic Index

The glycemic index is a numerical system that ranks carbohydrate-rich foods by how much they raise blood glucose levels compared to glucose (or white bread). This numerical system ranks foods on a scale of 0 to 100, according to their effect on blood glucose (sugar) levels. Foods that raise your blood glucose level quickly have a higher glycemic index ranking than foods that raise your blood glucose level more slowly. The Anti-Alzheimer's Diet encourages you to eat foods low on the glycemic index (see list on pages 82–83) to keep levels of blood glucose balanced and to prevent insulin surges.

Foods that are low on the glycemic index (fruits, vegetables low in starch, low-fat dairy, low-fat protein, legumes, nuts, seeds, and unrefined whole grains) will release slowly into your blood, helping to keep your blood sugar levels stable. Foods high on the glycemic index (white pasta, white bread, white rice, white potatoes, refined grains, sugary desserts) will cause blood-glucose levels to rise rapidly as there's a high response of insulin. Insulin then works fast to deposit the excess blood sugar into your muscle cells in the form of glycogen. When all your glycogen stores are full, the rest of the blood sugar is stored as fat. When you eat foods high on the glycemic index, it causes blood sugar to spike and then quickly fall to subnormal levels. This is why you feel energetic and then exhausted and starving after eating a meal high on the glycemic index.

If you are insulin resistant, eating foods high on the glycemic index may result in making you hungrier, causing you to gain weight and increasing your risk of cardiovascular disease, macular degeneration, and type 2 diabetes. Foods pure in protein and fat (such as lean chicken, fish, egg whites) have a low glycemic index, while a baked potato has a glycemic index of 85.

Lean protein, which is low on the glycemic index, helps to regulate the critical hormonal insulin-glucagon balance. As insulin slowly breaks down the protein you eat into glucose, glucagon is excreted by the pancreas. In addition, glucagon doesn't turn excess glucose to fat as insulin does. In fact, glucagon can convert fat to glucose while using up stores of fat. In doing this, it burns far more calories. Imagine eating succulent grilled salmon with a plate full of kale or other greens and actually losing fat and pounds at the same time!

Many studies now link memory impairment to type 2 diabetes, which happens when the body can no longer keep blood insulin levels controlled. In the Women's Health Study, scientists found that diabetes has an adverse effect on memory—almost like being ten to fifteen years older. *Controlling blood sugar is the key.* You can do that by eating foods that are low in sugar and low on the glycemic index.

I've given you lists of foods considered low on the glycemic index. It's important to note that some combinations of food can actually change where the food stands on the glycemic index. If a carbohydrate has fiber, it helps to decrease the glycemic load of the meal. I always tell my patients that a slice of whole grain bread has a lower glycemic load when they eat it with natural peanut butter. Why? Because the fat and protein of the peanut butter (low on the glycemic index) bring down the glycemic load of the whole grain bread.

SELECTING LOW GLYCEMIC INDEX FOODS
(less than a 55 on the GI)

Based on 50 grams of digestible (available) carbohydrate in a food and then measuring the effect of the food on blood glucose levels over a two-hour period.

Yogurt, low-fat (unsweetened)	14	Celery	15
Artichoke	15	Cucumber	15
Asparagus	15	Eggplant	15
Broccoli	15	Green beans	15
Cauliflower	15	Lettuce, all varieties	15

Yogurt, low-fat (artificially		Apples	38
sweetened)	15	Haricot beans, boiled	38
Peanuts	15	Pears	38
Peppers, all varieties	15	Tomato soup, canned	38
Snow peas	15	Carrots, cooked	39
Spinach	15	Plums	39
Tomatoes	15	Ravioli, meat-filled	39
Young summer squash	15	Black-eyed peas	41
Zucchini	15	Wheat kernels	41
Soybeans, boiled	16	All-Bran cereal	42
Cherries	22	Chickpeas, canned	42
Peas, dried	22	Peaches	42
Grapefruit	25	Lentil soup, canned	44
Pearl barley	25	Oranges	44
Milk, whole	27	Carrot juice	45
Spaghetti, protein enriched	27	Grapes	46
Kidney beans, boiled	29	Baked beans, canned	48
Lentils, green, boiled	29	Bread, multi-grain	48
Soy milk	30	Rice, parboiled	48
Apricots (dried)	31	Barley, cracked	50
Milk, fat-free	32	Whole grain	50
Milk, low-fat	32	Yam	51
Chickpeas	33	Kidney beans, canned	52
Milk, semi-skimmed	34	Lentils, green, canned	52
Rye	34	Kiwi	53
Vermicelli	35	Banana	54
Spaghetti, whole wheat	37	Sweet potato	54

Eat Antioxidants to Block Brain Disease

We have substantial evidence that implicates oxygen-free radicals as mediators of inflammation and/or tissue destruction in many types of degenerative disorders. Free radicals are molecules with unstable by-products of oxidation, the chemical process that causes iron to rust and a peeled apple or banana to turn brown. In the body, free radicals are like scavengers that cause similar deterioration, as they eat away at cell membranes and make cells vulnerable to decay and pathogens. These free radicals damage DNA and mitochondria, the basic building blocks of all tissues, and leave in their path many serious health problems, including Alzheimer's disease.

There's no known way to stop free radicals completely. And many recent studies indicate the increasing enormity of the problem with each cell in the body and brain being bombarded by these scavenger radicals, up to a thousand times per day. It is a constant battle.

We can reduce the free radicals' destructive effects on the body, supplementing the body's own defenses (anabolic hormones), by eating foods high in powerful antioxidants—plant chemicals that scavenge and neutralize the free radicals, converting them to harmless molecules. As antioxidants in foods intercept free radicals in the body, they protect cells from the oxidative damage that leads to aging and also to Alzheimer's disease. Examples of antioxidants include vitamin C, vitamin E, selenium, the carotenoids, and the flavonoids. One National Institutes of Health study of 3,718 Chicago residents, aged sixty-five and older, found that compared to people who consumed less than one serving of vegetables a day, people who ate at least 2.8 servings of vegetables a day saw their rate of cognitive change slow down by 40 percent.

Antioxidants Improve Cognitive Function

Interestingly, comprehensive studies show that levels of the antioxidants vitamins A and E and the carotenoids (including beta-carotene) are very low in people with Alzheimer's disease. To the contrary, a review of some long-term studies from Switzerland that spanned 1971 to 1993 and involved 442 subjects aged sixty-five to ninety-four show that higher ascorbic acid and beta-carotene plasma levels are associated with better performance in terms of memory and recall. Laboratory studies in animals also support evidence that chronic antioxidant treatment can improve cognitive function during aging.

If free-radical damage causes aging, then ingesting antioxidants in high enough quantities should be able to slow aging. Ironically, you don't have to eat *tons* of foods high in antioxidants to slow down brain aging! For instance, studies have shown that eating just ¾ cup of blueberries per day can turn back the clock dramatically. (I will explain why blueberries improve cognitive function later in this chapter.)

We think the antioxidants vitamins C and E reduce damage caused by beta-amyloid. Although we're not totally sure how these antioxidants work to decrease the risk of Alzheimer's, we do know that vitamin C cleans up free radicals, preventing them from damaging DNA, helps control inflammation, aids in wound healing, and wards off infection. Vitamin E is important for the maintenance of cell membranes, and many metabolic processes in the body are dependent upon healthy cell membranes—including the recuperation and growth of muscle cells. In scientific studies, those participants with the highest vitamin E intakes were more than 40 percent less likely to develop Alzheimer's disease; those with the highest vitamin C intake were more than 30 percent less likely to develop Alzheimer's. In the Rotterdam Study, a population-based study of 7,983 people aged fifty-five or older conducted in the Netherlands, researchers found that over a six-year period, those individuals who had a high dietary intake of vitamin C and vitamin E had a lower risk of Alzheimer's disease.

Studies also show that the catechins from green tea help protect brain cells. Findings suggest that green tea or green tea extract may be useful in protecting humans from senile disorders (including Alzheimer's). Green, white, and black teas are naturally rich sources of antioxidant flavonoids. Flavonoids, plant-derived antioxidants, may benefit the heart and brain by preventing low-density (LDL) oxidation, reducing inflammation, improving endothelial function, and inhibiting platelet aggregation. Some recent studies indicate that the antioxidants in tea are more powerful than the antioxidants found in many fruits and vegetables. This could mean that sipping on hot or iced green tea throughout the day may boost a bigger, healthier brain than eating platefuls of spinach! (*Brain Buster: Findings show that adding milk to hot tea reduces the influence of antioxidants on the body.*)

ANTI-ALZHEIMER'S DIET ANTIOXIDANTS

Apples
Apricots
Bananas
Beets
Bell peppers
Blackberries
Blueberries
Broccoli
Brussels sprouts
Cabbage
Carrots
Cauliflower
Citrus fruits
Eggplant
Grapefruit, pink
Kale, collards, or other greens
Kiwis

Kohlrabi
Mangos
Melons
Onions
Oranges or orange juice
Pineapples
Plums
Raisins
Raspberries
Red grapes
Romaine or other leaf lettuce
Spinach
Strawberries
Sweet potatoes
Tea (green tea is best)
Tomatoes or tomato juice

OMEGA-3 FATTY ACIDS STOP OXIDATIVE DAMAGE

Omega-3 is found in fish and fish oils. Some recent medical studies have revealed that deficiencies of micronutrients may result in an increase in pro-inflammatory markers in the body, altered immune response, and cell destruction. Scientific findings now indicate that inflammation in the body triggers cellular destruction in the brain, either directly or indirectly—even when the inflammation is mild.

Evidence continues to support the immune-modulating properties of omega-3 fatty acids in fish oil, eicosapentaenoic acid (EPA) and docosahexaenoic acid (DHA). Not only do omega-3 fatty acids have innumerable anti-inflammatory, anti-cancer, and anti-hypertensive effects, they are crucial to the prevention and treatment of Alzheimer's.

DHA Is Necessary for Cognitive Function

Docosahexaenoic acid or DHA is the most prominent fatty acid in the brain and is necessary for vision and cognitive function. Study after study confirms that the brains of Alzheimer's patients have a lower content of DHA in the gray matter of the frontal lobe and hippocampus than do the brains of persons without Alzheimer's disease. DHA is definitely one fatty acid you must include in your daily diet.

Findings from the Framingham Heart Study, which tracked about 900 healthy older men and women living in Boston for about nine years, showed that persons with high plasma DHA levels had a significantly lower risk of developing Alzheimer's than did those with lower levels. In fact, older adults who ate the equivalent of three servings of fish weekly had about half the chance of Alzheimer's when compared to adults who ate fish infrequently. In the Zutphen Elderly Study in the Netherlands, which followed 210 older men for five years, the men who consumed fish had less cognitive decline than men who did not consume fish. Similar results came from the Rotterdam Study in the Netherlands, where researchers found that eating fish just once a week was associated with a 60 percent reduction *in the risk* of developing Alzheimer's disease.

Another recent study found that a diet rich in the fatty acid DHA might interfere with the abnormal clumping of beta-amyloid and tau. When these two proteins clump in the brain, lesions known as plaques and tangles form. Researchers from this study believed that DHA might confer its benefits by lowering levels of an enzyme needed to generate beta-amyloid. Likewise, findings presented at the American Psychosomatic Society's Annual Meeting in 2007 in Budapest, Hungary, reported that a higher intake of omega-3 fatty acids is associated with greater volume in areas of the brain related to mood and behavior.

Omega-3s Help the Heart and Brain

Numerous findings on fish or fish oil supplements report improved triglycerides, high-density lipoprotein cholesterol, platelet function, endothelial and vascular function, blood pressure, cardiac excitability, measures of oxidative stress, pro- and anti-inflammatory cytokines and immune function. Both EPA and DHA are constituents of the membranes of all cells in the body and are precursors of locally produced eicosanoids. Eicosanoids are powerful hormones that include prostaglandins, thromboxanes, and leukotrienes. They are derived from long-chain essential fatty acids that affect the synthesis of every other hormone in the body and collectively mediate almost every component of the inflammatory response. The latest findings on those who eat fish or take fish oil supplements report improved blood lipid profiles and also a reduction in blood pressure, arrhythmias, and coagulability, and improvement in endothelial function. These are all important for reducing the risk of heart disease—and brain disease.

Fish, Mercury, and Omega-3 Fatty Acids: Help or Harm?

Some recent reports from the Food and Drug Administration and the Environmental Protection Agency warn consumers about eating fish because of the high levels of mercury, PCBs (polychlorinated biphenyls), dioxins, and other environmental contaminants. These reports have caused many people to be wary of fish consumption for fear of mercury toxicity. Yet we have tens of studies proving that fish contains the omega-3 fatty acids that are essential for heart and brain health, especially with aging.

According to the American Heart Association (AHA), which recommends eating fish at least twice a week for the prevention of cardiovascular disease, the benefits and risks of eating fish vary depending on your age and stage of life. As an example, young children, pregnant women, and women who are nursing usually have low risk of cardiovascular disease but are at higher risk from exposure to excessive mercury from fish. The AHA recommends that these groups avoid potentially contaminated fish. Yet for middle-aged and older men, and women after menopause, the AHA says the benefits of eating fish far outweigh the risks, which are explained in the guidelines of both the Food and Drug Administration and the Environmental Protection Agency.

The Anti-Alzheimer's Prescription

If you are a middle-aged or older man, or a woman after menopause, I encourage you to eat at least two servings a week of the following fish to increase your intake of omega-3 fatty acids so necessary for reducing inflammation in the body.

To help minimize the potentially adverse effects of environmental pollutants, be sure you eat a different choice of fish each week from the following list:

FISH HIGH IN OMEGA-3s

Anchovies	Salmon (wild)
Bluefish	Sardines
Capeline	Shad
Dogfish	Sturgeon
Herring	Tuna (canned or fresh)
Mackerel	Whitefish

N.B.: Wild salmon has high levels of omega-3s fed naturally in the environment. Atlantic or farmed salmon are grown on omega-6 grains and contain low levels of omega-3s and are therefore not beneficial. Always check when you order salmon at a restaurant or at a store if it is farm-fed. A hint is that it is much cheaper in price and poor in omega-3 value. Mercury is highest in shellfish, swordfish, shark, and some tunas. It is lowest in freshwater fish.

HOLD THE BEEF

I've talked a lot about eating fruits, vegetables, and fish high in good fats. What's the alternative? The other end of the nutrition/wellness spectrum reveals that a diet high in cholesterol and fatty acids may actually *increase* the risk of Alzheimer's disease. In large studies, researchers have found a *definitive link* between cognitive decline and Alzheimer's in those populations with a higher dietary intake of saturated fats, trans fats, and cholesterol. Aim to hold the beef in your diet and focus on plant-based foods, lean protein, low-fat dairy, and fish. If you must eat meat, buffalo, lean pork, and skinless poultry are preferable.

Folic Acid Improves Cognitive Function

In chapter 2, I discussed some new research that showed B-vitamin supplements, including folic acid, decrease the risk of cardiovascular disease by lowering levels of the amino acid homocysteine, which is thought to damage the inner lining of arteries. Aware for some time that high levels of homocysteine may also be associated with poor cognitive function, researchers have theorized that reducing homocysteine with folic acid may also reduce the chance of Alzheimer's, or, better still, increase cognitive function.

While preliminary, findings published in the journal *The Lancet* indicate that older adults who took folic acid supplementation over three years had improved global cognitive function with better memory and ability to process information—problems normally associated with aging. The study was part of the Folic Acid and Carotid Intima-media Thickness (FACIT) trial, and at the start, both the folic acid group and the placebo group had similar scores on a battery of tests for cognition, memory, sensorimotor speed, complex speed, information-processing speed, and word fluency. However, after three years, researchers using a battery of tests found that the change in memory and sensorimotor speed were significantly better in those participants who supplemented their daily diet with 800 micrograms of folic acid than those in the placebo group. Sensorimotor speed measures basic speed and shows direct stimulus-response connections, whereas complex speed measures time needed for higher-order information processing. The three-year change in cognitive function in terms of information-processing speed was also significantly better in the folic acid supplementation group than in the placebo group.

Study results also determined that while folate concentrations *increased* by 576 percent in the participants taking folic acid, plasma total homocysteine concentrations *decreased* by 26 percent. Realizing that high levels of homocysteine are also linked to damage to the hippocampus—the area of the brain that's important for memory formation—the researchers suggested that folic acid might simultaneously affect memory and sensorimotor speed.

Since both participants and researchers were unaware of the treatment, it makes the outcome of the cognitive function tests difficult to ignore, especially with participants demonstrating the following results:

1. 4.7 years younger for memory
2. 1.7 years younger for sensorimotor speed
3. 2.1 years younger for information-processing speed
4. 1.5 years younger for global cognitive function

Perhaps the most significant finding in the FACIT trial was the result for delayed memory recall (a fifteen-word learning test), particularly as most people associate memory loss with aging. After three years of supplementation with 800 micrograms of folic acid, the study participants improved their performance to be equal to a person six to nine years younger.

Most patients ask me if they need to take folic acid supplements, and here's what I tell them: *If you eat plenty of vegetables, you probably have an abundance of folate in your body.* However, according to the Institute of Medicine, the body absorbs only about 50 percent of food folate, while approximately 100 percent of the folic acid in a vitamin supplement is absorbed. Also, if you have a diet high in animal protein and low in fruits and leafy vegetables, you may have lower levels of folate and higher serum homocysteine levels.

In addition, because of poor diet or interactions with medications, many elderly people often have low levels of folate, which may correlate with high serum homocysteine and cognitive decline. Deficiency of folate is linked to a wide variety of nervous system problems, including general mental fatigue, non-senile dementia, depression, nervous-system problems in the hands and feet, irritability, forgetfulness, confusion, and insomnia—all problems which often coexist in older adults.

FOODS HIGH IN FOLATE

Asparagus
Beans (chickpeas, black beans,
  kidney beans, lima beans)
Cantaloupe
Citrus fruits
Eggs
Fortified grains

Green leafy vegetables (spinach,
  greens)
Legumes
Melons
Nutritional yeast
Strawberries

PUTTING THE PROGRAM INTO ACTION:
START THE ANTI-ALZHEIMER'S DIET

I know you're busy, and the last thing you want is another list of Dieting Dos and Don'ts. So I've simplified the Anti-Alzheimer's Diet to these three basic tips:

Tip #1: Do a Kitchen Overhaul. Get rid of all highly processed foods, foods high in sodium (salt, MSG), and foods high in bad fats, sugar, and white flour. I will explain more on page 93.

Tip #2: Go Shopping. Fill your refrigerator and cupboards with the Golden Dozen (page 103), along with the other recommended foods in this chapter and in the Shopping List on page 253. These brain-boosting foods are *proven* to help keep insulin levels balanced, if eaten in the proper proportions, and block inflammation and harmful free radicals in the brain.

Tip #3: Follow the Harmonic Method of Eating (the ⅓ Proportion Rule). Use a ⅓–⅓–⅓ distribution of calories from good fat, lean protein, and complex carbohydrates throughout the day. In addition, I have included the 28-day Menu for the Anti-Alzheimer's Diet in the appendix (page 257). Here, you'll also find a helpful, itemized Anti-Alzheimer's Shopping List with specific name brands that I recommend and forty superb-tasting recipes that are healthy for your brain.

Although I'm a strong proponent of eating whole foods to get all the available nutrients for the body, I will also recommend some necessary Anti-Alzheimer's natural dietary supplements (page 107) that you might consider taking to ensure that you get all the brain-boosting benefits you need. There is increasing scientific evidence that these supplements are crucial for preventing Alzheimer's disease.

TIP #1: DO A KITCHEN OVERHAUL.

Clean out your kitchen before you start the Anti-Alzheimer's Diet. Go through your cabinets, the pantry, the refrigerator and freezer, and throw out (or give away) foods that are *not* recommended on the Anti-Alzheimer's Diet. This includes foods that are high in saturated fats, partially hydrogenated or trans fats, and highly processed foods. You'll also toss foods high in sugar, white flour, and artificial additives. On the other hand, keep products that contain whole grains, lean protein, good fats, and complex carbohydrates—all are broken down slowly in the body, allowing you to stay in hormonal harmony with insulin in control.

Learn How to Read the Nutrition Facts on Food Labels

As you do the Kitchen Overhaul, it's important to read the Nutrition Facts on food labels to get pertinent information about calories, fats, protein, fiber, specific vitamins and minerals, and ingredients. This information is located on the outside of the package and is easy to read.

Check Out the Serving Size

Starting with the top of the Nutrition Facts, you will read the serving size (such as ½ cup, 5 crackers, or 10 chips) and servings per container (such as 2, 4, 6). The label then shows the amount of calories per serving and the amount of calories from fat. These numbers are important, especially if you aim to eat a diet lower in calories and fat. For example, having five whole grain crackers at eighty calories per serving is not awful for a snack. But who eats just five crackers? If you had fifteen crackers, you'd consume 240 calories—which is probably too many, especially if you're watching your weight.

You'll notice different units of measurement on food labels. Many of the nutrients are measured in grams or "g," while others are measured in milligrams or "mg." Some information is given in percentages (%).

Read About Fats and Other Nutrients

Along with calories per serving and calories from fat, the Nutrition Facts gives you the amount of total fat. It then breaks the total fat number down into saturated fat and trans fat—the unhealthy fats not recommended on the Anti-Alzheimer's Diet because they increase the risk of certain diseases. The total fat number is also broken down into polyunsaturated fat and monounsaturated fat, which are more beneficial to your health. Let's look at what these fatty terms mean:

1. **Bad Fat: Cholesterol** is found mainly in meat, poultry, fish, and dairy products. The food cholesterol increases cholesterol in your blood and adds to the risk of heart disease.

2. **Bad Fat: Saturated fat** comes from foods that are solid at room temperature, including animal sources, dairy products, and some oils. Saturated fat is found in red meat, butter, cheeses, luncheon meats, cocoa butter, coconut oil, palm oil, and cream. An excess of saturated fats raises the cholesterol level in the blood.

3. **Bad Fat: Trans fats** are formed during the process of hydrogenation. These fats are in foods such as vegetable shortening, some margarines, crackers, candies, cookies, snack foods, fried foods, baked goods, salad dressings, and other processed foods. Eating too many trans fats raises the cholesterol level in the blood.

4. **Good Fat: Brain-boosting polyunsaturated fat** includes both omega-3 and omega-6 fatty acids and comes from foods that are

liquid or soft at room temperature, including plant foods, nuts, seeds, some plant oils (sunflower, corn, soybean) and some seafood (herring, salmon, mackerel, halibut—all high in omega-3 polyunsaturated fatty acids), and soybeans. Polyunsaturated fat is necessary to protect the body against illness.

5. **Good Fat: Monounsaturated fat** comes from plant foods that are liquid at room temperature, including olives and olive oil, along with canola oil, peanuts, and avocados. New research suggests that these fats help to reduce your risk of heart disease (what is good for your heart . . . is good for your brain).

After listing the fats, the carbohydrates, dietary fiber, sugars, and protein are listed next on the food label. These items are followed by specific nutrients in the food product such as vitamin A, vitamin C, calcium, and iron. Lastly, the food label lists the ingredients in the product.

What Do Daily Values Mean?

To the right side of the Nutrition Facts are the Daily Value percentages. The Percent (%) Daily Value indicates how much of a specific nutrient one serving of food contains compared to recommendations for the entire day. The percentages next to each nutrient—fat, sodium, fiber, protein—guide you in determining if a food is "high" or "low" in specific nutrients, depending on the Daily Value recommendations. Five percent or less is considered to be low; 20 percent or higher is high. For example, the Dietary Fiber is 0 percent or low in Ritz Crackers. In All-Bran Multi-Grain Crackers, the Dietary Fiber is 5 grams or 20 percent of your Daily Value, which is high—a much better choice, in my opinion.

Avoiding Marketing Hype

As you become used to reading food products' Nutrition Facts, you'll realize that some manufacturers try to fool consumers. Some packages say "all natural," but if they are excessively high in sugar or saturated fat, all natural means nothing! If a food label says "low-fat," read the Nutrition Facts to see if it's a healthy choice. Many times a low-fat food is still high in sugar or calories.

Boosting Key Nutrients

It's important to choose food that is nutrient dense, which means food that delivers a complete nutritional punch—filled with substantial levels of vitamins and minerals, with few calories and limited saturated and trans fats, and low in cholesterol, sodium, and sugar.

Limit: Total Fat, Saturated Fat, Trans Fat, Cholesterol, and Sodium. Eating too much of these may increase your risk of heart disease, high blood pressure, and some types of cancer.

Increase: Good Fats, Dietary Fiber, Vitamin A, Vitamin C, Calcium, and Iron. Eating plenty of these key nutrients can boost your immune function and overall health. Plenty of fiber is important to promote healthy bowel function, while calcium builds strong bones and prevents fractures. Vitamins A and C are important for staying well, preventing infection, and reducing the risk of diseases.

What I Tell My Patients About Ingredients Listed in Nutritional Facts

Here's what I tell my patients: If you read the Nutrition Facts and you don't understand the ingredients . . . *toss it.* If you're at the grocery store and you read the Nutrition Facts and don't understand some ingredients or terms, *do not buy it.* The best way to "know" you're getting safe and healthful foods is to purchase foods that have *few ingredients* and include whole foods (vegetables, fruits, whole grains, legumes, fresh meats and fish, eggs, soy, low-fat milk, and yogurt). If the list of ingredients is longer than three to five foods, put it back on the shelf. I recommend shopping the perimeter of the grocery store, because that's where the whole foods are available for purchase.

Get rid of the following foods during your Kitchen Overhaul:

- *Highly processed foods*

You'll notice on the Nutrition Facts label that these foods have many ingredients—some that you've never heard of before. That's not a good sign when it comes to your health! Many ingredients? Toss the food. Several ingredients? Check them out, and if they are whole foods (whole grains, soy products, walnuts, milk, eggs), the product may be healthy.

• *Foods high in corn syrup, high-fructose corn syrup, and sugar*

Be sure to check the Nutrition Facts on your bread package and cracker box, too. If it has added high-fructose corn syrup (a liquid sweetener made from dextrose or glucose from cornstarch) or any of the following different names for "sugar," you might want to be aware and reduce your intake. Whereas high-fructose corn syrup, crystalline fructose (made from cornstarch if used as an ingredient in foods), and all sugars are generally safe, according to the FDA, added sugars to foods and juices usually decrease levels of more essential nutrients found in whole foods such as fresh fruits and vegetables. Here are some additional names for "sugar" to watch out for in all food labels:

• Brown sugar	• Maltose
• Concentrated fruit juice	• Maple syrup
• Demerara sugar	• Molasses
• Dextrose	• Panocha
• Free-flowing brown sugars	• Powdered or
• Fructose	confectioners' sugar
• Galactose	• Raw sugar
• Glucose	• Rice syrup
• Honey	• Sucrose
• Invert sugar	• Sugar (granulated)
• Lactose	• Syrup
• Malt	• Treacle
• Maltodextrin	• Turbinado sugar

• *Foods high in white flour*

Whether it says "enriched" or not, white flour is a fast-burning refined carbohydrate that causes insulin levels to spike. When they remove the whole grain to manufacture white flour, they strip out all B vitamins. Instead, select foods that contain grains and seeds like wheat, brown rice, barley, oats, millet, semolina, and whole-wheat flour.

- *Foods high in saturated fats, trans fats, or partially hydrogenated oils*

These fats increase lipids and triglycerides in the body, which increase your risk of Alzheimer's disease. If the ingredient label has any of these types of fats listed, toss it now.

- *Foods high in sodium, salt, and MSG*

Read your nutrition label to get the amount of sodium in a product. I recommend that patients stay below 130 milligrams of sodium per serving if they want to watch sodium in the diet. If there's MSG in a product and it says sodium on the label, this is being used to account for the sodium in the MSG.

LOW-CALORIE SWEETENERS AT A GLANCE

Sweetener	Caloric Value	Date Approved	Sweeter Than Sugar	Brand Name(s)
Acesulfame-K	0	1988	200x	Sunett, Sweet One
Aspartame	4	1981	180x	NutraSweet, Equal, others
Neotame	0	2002	7,000x	n/a
Saccharin	0	Years prior to 1958	300x	Sweet'N Low, Sweet Twin, Sugar Twin, others
Sucralose	0	1998	600x	Splenda

Source: *Comprehensive Reviews in Food Science and Food Safety*, IFT, 2006

- *Foods filled with sugar substitutes*

If you have diabetes and your doctor has recommended sugar substitutes, then follow your doctor's advice. If you don't have diabetes, you may want to remove the above sweeteners from your kitchen pantry. While the FDA has approved all of these, there are studies that question the safety, although the jury is still out.

TIP #2: GO SHOPPING.

Fill your kitchen with the Golden Dozen and other brain-boosting foods. Using the Anti-Alzheimer's Shopping List on page 253, go to the supermarket and purchase an array of fresh fruits, vegetables, legumes, whole grains, lean meats, fish, eggs, low-fat dairy, soy products, and nuts and seeds. I find that if you shop the *perimeter of the grocery store*, it's much easier to select the healthiest items that fit in the diet plan.

Many patients ask how they can know if a bread or cracker is a whole grain. I tell them to look for the word *whole* on the Nutrition Facts label. If the ingredient is 100 percent whole grain, oat bran, barley, or rye, these are acceptable. If *whole grain* is not listed, it means the germ is refined and not a whole grain product. When food manufacturers remove the whole grain and refine the bread (so it sticks to the roof of your mouth!), they strip out all the B vitamins. B vitamins are crucial for heart and brain health. Accept only whole grain products.

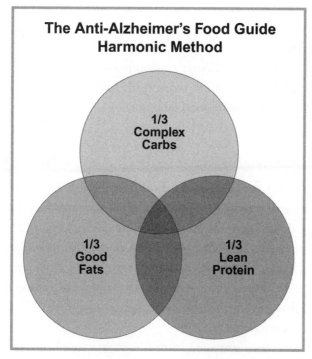

Figure 4.1 The Anti-Alzheimer's Food Guide: Harmonic Method

Tip #3: Follow the Harmonic Method of Eating
(the 1/3 Proportion Rule).

Select foods from the Anti-Alzheimer's Diet Shopping List (see Appendix A) and use the ⅓–⅓–⅓ distribution of calories from good fat, carbohydrates, and protein throughout the day. This proportion is easy to remember and crucial to maintain one's hormonal balance of insulin and glucagons.

⅓ Complex Carbohydrates
One-third of your daily calorie consumption should consist of complex carbohydrates (fruits, vegetables, and whole grains that are low in starch and high in fiber).

⅓ Lean Protein
One-third of your daily calorie consumption should consist of very lean protein, including fish, lean meats, poultry, eggs, soy products, nuts and seeds, and low-fat dairy.

⅓ Good Fats
One-third of your daily calorie consumption should consist of "good" fats. Olive oil, grape seed oil, peanut oil, flaxseed oil, olives, avocados, and nuts are all filled with good fats.

Lose Weight and Reduce the Chance of Alzheimer's

A key risk factor for Alzheimer's disease is obesity. We know that people who are overweight or obese in their forties have a greater risk of developing Alzheimer's disease later in life. Research suggests that visceral fat present in large amounts in the midsection of the body (what we call belly fat) might be a major source of inflammatory molecules. The resulting inflammation is suspected to play an important role in diseases such as insulin resistance, hypertension, type 2 diabetes, and possibly Alzheimer's.

The daily menus provided in this book are based on the consump-

FIBER AND THE ANTI-ALZHEIMER'S DIET

Vegetables, fruits, legumes, whole grains, nuts, and seeds are all high in fiber. On the Anti-Alzheimer's Diet, you should aim for 30 to 35 grams of fiber daily, which will be easy with the choices I've given you. Fiber helps to regulate blood glucose levels and keep you full. Insoluble fiber found in wheat and whole grains, apple peel, celery, carrots, greens, broccoli, green beans, and squash attracts water and swells up, adding bulk to the stool. This type of fiber is necessary to keep your bowels regular and also helps with weight control. Soluble fiber dissolves in water and slows the absorption of sugars, fat, and cholesterol in the intestines. Soluble fiber can help to lower abnormal lipids, especially when your diet is low in saturated fat and cholesterol. Foods high in soluble fiber include oat bran, oatmeal, oranges, apples, carrots, dried beans and legumes, peas, rice bran, barley, citrus fruits, strawberries, and apple pulp. They help steady your blood sugar, reducing the spikes in insulin.

tion of 2,000 calories per day. However, some may need additional calories and some may need less depending on your personal caloric needs. The menus are based on a ⅓–⅓–⅓ distribution of calories from good fat, carbohydrates, and protein consumption throughout the day. It is important to maintain this distribution on a daily basis in order to regulate blood sugar levels. Therefore, if your calorie needs are less than 2,000 calories per day, it is advised that you consume smaller portion sizes of each of the foods listed in the daily menu. For example, if lunch provides chicken, rice, and beans, do not cut out just the beans. Try to cut back on the portion of chicken, rice, and beans evenly. The same concept applies if you need to consume more that 2,000 calories per day in order to maintain your weight. It may be necessary that you consult a registered dietitian to help you determine your personal calorie needs for weight maintenance or weight loss. In order for you to find a dietitian in your area, you can use the American Dietetic Association Web site (www.eat right.org).

More Anti-Alzheimer's Diet tips include:

- Exercise as a family.
- Eliminate sugary beverages in the household.
- Promote the consumption of fruits and vegetables by having them cleaned and readily available in the refrigerator or on the kitchen table.
- If you must have a sweet snack . . . eat only one "treat" each day (make sure it is no more than 150 calories to limit sugar intake).
- Cook with nonstick cookware and olive oil cooking sprays to cut back on oils and fats in the diet.
- Be a role model. . . . Eat healthy and exercise in front of your children or grandchildren.

The Anti-Alzheimer's "diet" is an eating plan for life, and the menus provided will help guide you through your journey. It may not always be realistic that you will follow a 28-day meal plan repeatedly; however, when you feel you need to get back on track, the menus will be here to get you back on the right path. Use the menus for meal ideas and as a guide for appropriate serving sizes. To maintain a healthy diet, plan your meals for the week; make a grocery list and purchase what you can at the beginning of the week. Buy fresh foods, such as fish, fruits, and vegetables, as you need them throughout the week. Make extra portions for dinner and bring leftovers for lunch or use them for dinner another night of the week. . . . This will cut back on time spent in the kitchen!

Along with weight loss, you can lower your blood pressure, LDL ("bad") cholesterol, levels of C-reactive protein, and risk for metabolic syndrome—all health conditions that increase the chance of Alzheimer's.

AGE AND WEIGHT GAIN

There are a number of reasons why we get fatter with aging.
1. Our thyroid, responsible for our metabolic rate, decreases.
2. In the hypothalamus of the brain, leptin receptors are fewer as we age. Leptin is important for satiety, or fullness, which makes you less likely to feel hungry.
3. A decrease in body lean mass is very important because muscle burns more calories than fat by a factor of 4 to 1.
4. As cortisol increases with age, neuropeptide Y increases and this increases weight gain.

The Golden Dozen

With the help of my highly qualified registered dietitian, Lauren Brand, M.S., R.D., we selected the following foods to be part of the Golden Dozen. Learn why these foods are important and then use them in your daily meal and snack choices. Lauren has prepared incredible recipes (page 271) using these brain-boosting foods, as well as a 28-day Anti-Alzheimer's Diet Menu on page 257.

1. Berries

Berries, including blueberries, strawberries, raspberries, acai, cranberries, bilberries, dark red cherries, and blackberries, are filled with anthocyanins, special chemical components that give intense color to fruits and vegetables. It is thought that anthocyanins sweep out harmful free-radical molecules that trigger inflammation. In an animal study from Tufts, rodents given a blueberry-supplemented diet not only experienced improvements in neuronal and cognitive behavioral activity, but their motor behavior also benefited. Other studies show that antioxidant-filled berries help fight against aging problems, such as short-term memory loss. Dark blue and purple berries also are linked to a significant reversal in motor dysfunction that correlates with aging and dopamine deficiency.

Quick uses: Toss berries in dinner salads, sprinkle on yogurt as a topping, or eat for snacks.

2. Apples

Apples are loaded with quercetin, a bioflavonoid that modifies the inflammatory response by inhibiting the release of prostaglandins, an inflammatory compound. The latest scientific studies confirm that inflammation is a primary "brain killer" and destroys from within. Be sure to eat your apple with skin, because the vitamin content is reportedly higher in the skin than the flesh.

Quick uses: Bake, slice in garden salads, or eat anytime for snacks.

3. Fish

Fatty fish like salmon, sardines, mackerel, and tuna are a great source of omega-3 fatty acids that reduce inflammation. Fish reduce triglycerides and lower blood pressure, too. (If you are a woman of childbearing age, talk to your doctor about including fish in your diet.)

Quick uses: Broil salmon, serve tuna for lunch, snack on sardines and crackers.

4. Cruciferous vegetables

Cabbage, broccoli, cauliflower, and Brussels sprouts are cruciferous vegetables filled with phytochemicals (plant protectors) that are vital in keeping your brain healthy. Cruciferous vegetables are a major dietary source of isothiocyanates, special compounds in food that may protect against coronary artery disease and Alzheimer's. In a twenty-five-year study of 13,000 women, researchers at Harvard Medical School found that those women who ate high amounts of vegetables, particularly cruciferous vegetables such as cabbage, had less decline in memory as they aged.

Quick uses: Steam, sauté, mix in casseroles, or shred (raw) for coleslaw.

5. Nuts

Nuts—including walnuts, almonds, hazelnuts, peanuts, and cashews—are powerhouses of monounsaturated fat, a good fat that helps to reduce LDL (bad) cholesterol and makes nuts both heart and brain healthy. Nuts are high in vitamin E, a powerful, fat-soluble antioxidant that helps protect cell membranes from damage by free radicals and prevents the oxidation of LDL cholesterol. Nuts are not as high in calories as many think because 15 percent of the calories are not absorbed by the gastric tract due to their covering. The slow release of protein and fat by nuts increases the feeling of fullness and the excretion of glucagon and not insulin.

Quick uses: Add to rice dishes, toss in cold salads, or eat a few nuts at snack time.

6. Dairy (low-fat)

Studies show that a diet high in low-fat dairy products significantly lowers blood pressure—and hypertension and cardiovascular disease are key risk factors for Alzheimer's. In addition, yogurt, which is high in protein, contains the amino acid tyrosine that's necessary to manufacture the neurotransmitters dopamine and noradrenaline. Scientific studies show that chronic stress depletes the body of tyrosine. But supplementing with certain foods—like yogurt—can increase memory and boost alertness.

Quick uses: Drink fat-free or low-fat milk with meals, pour on cereal, use in cooking; eat yogurt or low-fat cheese for snacks. Try the Anti-Alzheimer's Brain-Boosting Berry Smoothie (page 273).

7. Greens

Greens such as kale, turnip, mustard, collard, spinach, and other leafy greens are loaded with antioxidants that nourish and defend body cells, including neurons. These powerful antioxidants prevent plaque buildup in the arteries, allowing for a strong blood flow to the brain.

Quick uses: Steam, sauté, or chop for salads.

8. Dried beans or legumes

Dried beans or legumes—including lima beans, pink beans, navy beans, cranberry beans, red beans, black beans, white beans, pinto beans, great northern beans, chickpeas, and kidney beans—are high in folate, and folate is necessary for normal levels of homocysteine (high levels of serum homocysteine are associated with cognitive decline). An analysis of data from the Baltimore Longitudinal Study of Aging has revealed that those with higher intake of folate and vitamins B_6 and E had a lower risk of developing Alzheimer's.

Quick uses: Serve hot at mealtime, mix in casseroles, or toss in cold salads.

9. Soy

Soy products are excellent sources of plant protein that are high in polyunsaturated fats, fiber, vitamins, and minerals and low in saturated fat. Two components of the soybean, daidzein and genistein, have beneficial effects in cancer prevention and the prevention of heart disease. Eating soy foods helps lower LDL (bad) cholesterol that's associated with an increased risk of heart disease—and brain disease (including Alzheimer's).

Quick uses: Use soy milk on cereal or in baking; use soy-substitute crumbles instead of meat in recipes; use dried beans in recipes such as chili and baked beans.

10. Sweet Potatoes

Loaded with significant values of beta-carotene, vitamin C, vitamin B_6, folate, potassium, phosphorus, calcium, and fiber, sweet potatoes are associated with reducing the risk of cancer, suppressing tumor development, and even improving night vision. Because of the fiber content, sweet potatoes have a lower glycemic load than white potatoes.

Quick uses: Use in place of white potatoes; serve baked sweet potato fries.

11. Tomatoes

Preliminary research confirms that the potent antioxidant lycopene, which is prominent in tomatoes, may be more powerful than beta-carotene, alpha-carotene, and vitamin E. This antioxidant is associated with a better heart and memory and even protection against certain types of cancer. Studies show that heating tomatoes makes lycopene more easily absorbed and utilized by the body.

Quick uses: Stew them, add to soups and sauces, eat sliced on sandwiches or in salads.

12. Whole Grains

I urge you to select whole grain products whenever possible. Whole grains are low on the glycemic index (page 82) as it takes longer for them to digest, which slows down the conversion of starch to sugar. Whole grain products contain the word *whole,* not *enriched,* and include all three parts of a grain kernel: the bran, germ, and endosperm. Sprouted-wheat bread, some whole grain breakfast cereals, brown rice, barley, and oatmeal are all whole grain choices. Most whole grains contain from 1 to 4 grams of fiber per serving. I tell my patients that if their jaws are working out harder, chances are the grain is a true whole grain.

Quick uses: Use whole grain bread for sandwiches; have oatmeal for breakfast; add barley to casseroles.

ADD EGGS TO YOUR DIET

Eggs are rich in choline, which your body uses to produce the neurotransmitter acetylcholine. Low levels of acetylcholine are associated with Alzheimer's disease.

USE BRAIN-BOOSTING HERBS AND SPICES

Garlic, an antioxidant, is anti-atherosclerotic and can potentially help against stroke and Alzheimer's disease. Garlic has been found to lower blood pressure, protect blood vessels, and prevent plaque buildup in arteries, an important factor in preventing heart disease and stroke. Some new findings indicate that garlic is also neuroprotective, supporting both memory and new learning.

Curcumin, an antioxidant and component of the spice turmeric, has been shown to reduce oxidative damage and decrease beta-amyloid peptides in the brain by 43 to 50 percent. Studies show that elderly villagers in India who eat turmeric with every meal have the lowest rate of Alzheimer's disease in the world.

ANTI-ALZHEIMER'S SUPPLEMENTS YOU NEED

While I'd prefer that you get necessary nutrients from whole foods in your diet, you can consider the following supplements:

1. 1 multivitamin that does not contain iron or copper (If you are a woman of childbearing age and take iron supplements, ask your doctor about the best source.)
2. 800 micrograms of folic acid daily (This amount may be in your daily multivitamin.)
3. 1,000 mg (twice daily) of fish oil or flaxseed supplements (Note men with prostate problems should consult with their physician before taking flaxseed.)
4. Supplement with antioxidants:
   - Vitamin C: 2,000–4,000 mg
   - Vitamin E: 400–800 IU
   - Beta-carotene: 15,000–40,000 IU

- Selenium: 100–300 mcg
- Grape Seed Extract: 50–200 mg
- Coenzyme Q10: 300–1200 mg
- L-Lipoic Acid: 50–250 mg
- Green Tea Extract: 30–150 mg

5. 200–500 micrograms daily of chromium picolinate
6. 1 baby aspirin daily (for anti-inflammatory effects)
7. 1 glass daily of red wine. Findings from the Italian Longitudinal Study on Aging show that people with mild cognitive impairment who drink up to one glass of alcohol daily, mostly wine, developed Alzheimer's at an *85 percent slower pace* than people who never drank alcohol.
8. BriteShield, a supplement by Brite Age Corporation, has both turmeric and resveratrol, the active ingredient in wine. I recommend two BriteShield supplements daily, especially *if you don't wish to drink red wine or eat turmeric each day.*

If you cannot afford to purchase all of these supplements, the minimum I recommend to patients includes fish oil, folic acid, a multivitamin, and Coenzyme Q10, if you are taking a statin medication. (Coenzyme Q10 is very expensive, usually $25 or more for 60 pills of 150 mg each.) In addition, if you are taking statins it's important that you know there are some common side effects called myalgia or muscle aching. When you take statins and exercise, you may experience severe pain in your forearms, buttocks, or thighs probably related to the statins. This side effect can lead to a permanent autoimmune problem. To prevent serious problems, ask your doctor to do a laboratory test called a CPK. If this is elevated, you'll have to weigh the risks versus the benefits of the statin medication. There are good alternatives to your current treatment, so talk to your doctor about this problem and changing treatment.

Supplements: A Hard Pill to Swallow

The major reason supplements are not taken is they are "hard to swallow"—literally. I surveyed one hundred patients and more than 70 percent had some anxiety or problem with swallowing their supplement pills. The most common problems were that the pills were too large (and sometimes contained sharp edges), and the patient did not know the best way to swallow the pill.

I found the best-tolerated pills were small, roughly a centimeter in size, a little more than a third of an inch, and football shaped. Giant pills such as 1,000 milligram fish oil, flaxseed oil, or Coenzyme Q10 were often a problem.

Another problem I discovered is many patients thought that throwing their head back and swallowing (as birds do) would improve their ability to down these mega-pills. However, what may be good for birds is not for humans! Throwing your head back increases the likelihood of choking. Instead, keep your head in a neutral position, and place the pill in the middle of the tongue. While swallowing liquid, tucking your chin down about 50 percent of the way to the chest is optimum. Try sipping water out of the palm of your hand as an example of the correct head position for pill swallowing. Using a straw with the head tilted forward is best, as drinking from a cup often causes you to lift the chin and places our swallowing mechanism in a suboptimal position.

Other tips for swallowing pills include using a denser fluid such as milk or a smoothie. Some of my patients put their pills in yogurt; other patients coat them with a small amount of olive oil.

Fish oil pills are often a particular problem. These larger capsules usually cause a very distasteful regurgitation after consumption. This distasteful regurgitation can be avoided by keeping the fish oil bottle in the freezer, eating the capsules with food, or buying capsules that are enteric coded, which means they dissolve in the small intestine. Buying fish oil capsules that are 500 milligrams reduces the size of the capsule, too. Some brands, such as ProDHA, are flavored and of the proper size.

Attitude is important. Enjoy the supplements, but don't regard them as medicine. I advise taking them with eight to ten ounces of fluids. It helps stem the appetite.

Caution on Supplements and Herbal Remedies

Many of the supplements interact with other medications and have adverse affects by either enhancing or inhibiting the effects of medications given for heart, liver, or other illnesses you may have.

Copper is associated with an increased likelihood of Alzheimer's disease according to the Academy of Neurology. Vitamins should have less than 2.5 milligrams of copper per day.

Chondroitin sulfate, taken for joints, is noted to increase the risk of bleeding when taken with an NSAID (Motrin, Celebrex, and Daypro).

Flaxseed oil decreases the absorption of oral medications, especially blood thinner, and so should be taken separately.

Dehydroepiandrosterone increases insulin resistance and should be taken cautiously if you are insulin resistant.

Ginger interfaces with warfarin (Coumadin) and ginkgo biloba enhances bleeding tendencies.

Melatonin, often used for sleep problems, increases clotting and interferes with warfarin (Coumadin) anticlotting activity.

Saint-John's-wort can interfere with antidepressants and with amioderone, a cardiac anticoagulant, and increase the tendency to bleed.

Weight Loss Supplements

Care and caution must be paged to weight loss herbs and supplements. As an example, ephedra is well-known to cause a serious risk of heart attacks and irregular heartbeat. Other weight loss schemes, such as calcium supplements, bitter orange, and chitosan, do not work for anyone except the companies that sell them.

High-colonic enemas do not make you lose weight but they do clean out your wallet. Olestra, a fat substitute used to lower calories and fat in products such as potato chips, also reduces absorption of fat-soluble vitamins such as caratenoids. Finally, one added caution; nonsteroid anti-inflammatory drugs (NSAIDs) such as Advil, Motrin, Aleve, and Celebrex inhibit the anitplatelet effect of aspirin that's necessary to prevent heart attacks. So if you need aspirin for cardiovascular help, do not take it within fourteen hours of taking an NSAID.

KIDS & ALZHEIMER'S PREVENTION

According to pediatrician and author Dr. William Sears, kids are taught to choose healthy foods in elementary school and to extend these habits to adulthood. As a dad, I believe in the importance of a family dinner together as an opportunity to help kids form healthy work habits and social skills. At the family dinner, you can assist your child in planning and prioritizing homework and extracurricular activities. At the same time, you can listen to their heartfelt concerns and share stories on how you dealt with similar problems at their age.

One day, my son Vinnny, then eight, heard me telling my wife at dinner that some of my staff was overly sensitive, reacting to threats that didn't exist because they had an "inferiority complex." When I confronted them with this idea, they stopped it. The next evening before dinner, Vinny came up to me and told me that the class bully was picking on one of his friends. He said he stopped him by telling him he had an inferiority complex. "Dad, this really confused him and he left us all alone." Vinny, at the time, didn't know an inferiority complex from a superiority complex. (Today, Vinny is a well-known therapist.)

As a father and a physician, I know that kids listen to and mimic behaviors of caring adults in their lives. If you eat healthy foods in harmonic proportions, they will follow. The road to preventing Alzheimer's starts with the examples you give your children today.

STEP 2: BRAWN BOOSTERS

DAILY AEROBICS AND ANAEROBICS
FOR THE BODY AND MIND

When I was growing up, being healthy in body, mind, and spirit was more of a platitude than something people actually believed and lived. You may remember a time in the 1950s and 1960s when people believed that brains and brawn did not mix. In years past, the word "dumbbell" had a double meaning. Not only did it refer to the barbell, but it also referred to the weight lifter.

But in order to survive in the Italian Brooklyn neighborhood where I grew up, I began to lift weights. A "skinny Vinnie" was not going to make it in the testosterone-driven streets of Little Italy. In high school, I transformed into a closet weight lifter, as my Catholic high school had classes all day Saturday, which was the same day weight lifting contests were held. I remember conveniently coming down with stomachaches some Saturdays so I could miss school and compete in the North American and Junior Olympics Weightlifting Championships. In college, despite my being the number one prospect for the 1964 Olympic Weight Lifting Team, my classmates and professors at Seton Hall never knew of my secret brawn-building endeavors—I was a virtual Dr. Jekyll and Mr. Hyde. In the streets of Brooklyn, I was Mr. Hyde. At Seton Hall, I became Dr. Jekyll, fearful of being labeled a "dumbbell."

All I knew was that my grades and athletic performance improved greatly with my weight lifting, contrary to the belief that brains and brawn didn't mix.

Want a stronger brain? Building brawn builds a smarter brain. In fact, both aerobic (steps) and anaerobic (weight lifting or strengthening) exercises do the brain good! The older we get, the more important exercise becomes to balancing our hormonal symphony, augmenting the brain's growth and reserve, and reducing the effects of the sentinel risk factors (chronic stress and sleeplessness).

While there have been some scientific studies over the years that have supported my brain/brawn observation, nothing really has affected research as much as the recent findings by award-winning neurobiologist Fred Gage, Ph.D., at the Salk Institute in California. Gage's lab showed that, contrary to accepted beliefs, we humans are highly proficient at growing new nerve cells throughout our lifetime. In fact, small groups of immature nerve cells are found in the adult brain (a process called neurogenesis). Gage and his team showed that changes in behavior such as *exercising more* could actually affect neurogenesis and alter the brain's wiring—specifically in the hippocampus, the brain's seat of memory and learning and the first place destroyed by Alzheimer's. Gage's studies confirm what I have taught for years, that exercise is one of the essential ingredients to balance our hormones and neurotransmitters as we age.

EXERCISE MAKES THE BRAIN BIGGER . . . AND BETTER

With increasing knowledge of the brain and body physiology, some new concepts have surfaced on the nature of exercise benefiting the mind. Let's look at some exercise benefits that can help balance your hormones and neurotransmitters and also decrease your chances of Alzheimer's disease:

• **Exercise reduces stress.** Exercise is to stress as laughter is to depression. Stress is a sentinel risk factor that may cause binge eating, obesity, hypertension, diabetes, and depression. Studies show that stress provokes the release of adrenaline and cortisol at times twentyfold. When stress hormones like cortisol are overproduced, it can lead to depression, cognitive dysfunction, and difficulty in concentration and memory retrieval.

Exercise helps reduce the body's stress response by turning off the adrenal glands that produce the stress hormone cortisol. Exercise reduces blood glucose and insulin, the conductor of your hormonal symphony. With exercise, there is increased production of the all-important neurotransmitters dopamine and serotonin (the "feel good" and "calming" neurotransmitters and the essence of our willpower). See chapter 8.

Regular exercise also reduces the amount of circulating adrenaline in the body. In doing so, it relaxes blood vessels and results in a slower pulse rate and lower blood pressure. Exercise boosts the body's production of endorphins, pain-fighting molecules that may also be the reason for the "runner's high." Endorphins help to reduce anxiety, stress, and depression.

• *Exercise increases muscle mass and the body's metabolic rate.* To the chagrin of the couch potatoes, those trim, muscular men and women burn more calories by just looking good. The more muscle mass your body has, the more calories you burn per day. Because men have more muscle mass, they usually burn calories faster and lose weight more easily than women do. However, with resistance exercises, both men and women can experience weight loss because of the increase in muscle mass. While fat at rest burns two calories per pound, muscle at rest burns six to ten calories per pound. When in motion, muscles burn up to fifteen to twenty calories per pound. So get off the couch and start exercising. If you can't get to the gym, do isometrics.

• *Exercise stimulates anabolic hormone production.* Exercise triggers production of key anabolic hormones—thyroid, testosterone, estrogen, human growth hormone, and insulin-like growth factors (IGFs). These hormones, just by their presence, make us look and feel young. Yet these anabolic hormones also decline significantly with age unless you work out regularly. As an example, with normal aging, human growth hormone *decreases by 90 percent* from ages twenty to seventy. There is likewise a reduction in thyroid, tetosterone, and estrogen. However, regular exercise stimulates the production of human growth hormone—no matter what your age. Your metabolism also decreases with aging. Yet thyroid hormones are directly stimulated by exercise, resulting in an increase in our resting metabolic rate. A study comparing growth hormone injection versus a weight-lifting program in seniors showed similar results: Both groups had an increase of 14 percent of lean muscle mass. Therefore, exercise can save you a lot of money since growth hormone injections are pricey.

• *Exercise builds brain mass and capacity.* With regular exercise—including stretching, stepping, and strengthening—there is an increase in
brain agility and memory. In the Honolulu-Asia Aging Study, research-

ers assessed the distance walked per day in 2,257 physically capable men aged seventy-one to ninety-three years. After adjusting for age, researchers confirmed that the men who walked the least (less than one-quarter mile per day) experienced a higher risk of Alzheimer's compared with those men who walked more than two miles per day. In the Nurses' Health Study, findings show that long-term regular physical activity, including walking, is associated with significantly better cognitive function and less cognitive decline in older women.

In another study published in the *Journal of Gerontology*, researchers had one group of elderly volunteers do aerobic training three times a week for one hour while the other group of volunteers did nonaerobic stretching and toning. After three months, magnetic resonance imaging (MRIs) taken of the participants' brains showed that the aerobics group increased their brains' volume, reflected by new neurons and cells, and white matter (connections between neurons) in the frontal lobes, which contribute to attention and memory processing. Researchers concluded that the brain volumes of the aerobic exercisers were like people who were several years younger.

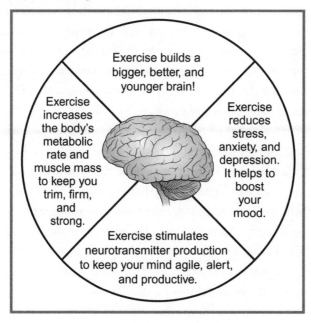

Figure 5.1 Benefits of Exercise

Hormonal Balance and Imbalance

To fully grasp the impact exercise has on the aging brain, it is important to understand the various hormones in the body and how they influence your memory, muscle mass, and weight. Hormones, chemical messengers produced in the body, are released into the blood by various endocrine glands such as the thyroid, pituitary, adrenal, pancreas, and gonads. Your hormonal system supports the full function of your body. If any hormone is released in amounts that are too high—or too low—it throws off every part of your mind and body, causing distress—or even illness. As an example, if thyroid hormones are too low, you will become severely obese, depressed, and mentally slow—even demented. If thyroid hormones are too high, you will become emaciated, anxious, manic, and an insomniac.

Anabolic Hormones

The anabolic hormones include thyroid, testosterone, estrogen, human growth hormone, and insulin-like growth factors (IGFs). As I discussed in chapter 3, I like to think of the anabolic hormones as having a positive influence on the body in concert with the neurotransmitters dopamine and serotonin, and the parasympathetic or vagal nervous system. In our hormonal symphony, the anabolic hormones with dopamine and serotonin are like the rich, deep tones of the piano and the soft humming strings of the violins and cellos.

Remember, the anabolic hormones work for you—*not against you.* They increase lean muscle mass and bone density and decrease fat stores. The result is an increase in your metabolic rate and weight loss and an increase in strength and endurance that results in increased attractiveness. Anabolic hormones also decrease the hormone insulin, stimulate the immune system, and decrease inflammation. This is big, for recent findings indicate that pro-inflammatory markers such as C-reactive protein (page 51) contribute to Alzheimer's disease. Studies show that exercise helps restore the body's neurochemical balance and triggers a positive emotional state.

Catabolic Hormones

The catabolic hormones include cortisol, progesterone, and DHEA. It's helpful to look at the catabolic hormones as the "enemy," with the exception of DHEA, as they contribute to the breakdown of muscle protein that's used for glucose (energy) synthesis. Catabolic hormones are associated with the sympathetic nervous system—the alarm side that's pumped by fight-or-flight adrenaline.

The catabolic hormones decrease muscle mass and bone density, and increase abdominal fat stores. Catabolic hormones result in the unhealthy and unappealing "couch potato body." Catabolic hormones also stimulate the sympathetic nervous system causing alarm, anxiety, high pulse rate, and hypertension. These hormones lead to depression when they become depleted along with many neurotransmitters that were being used for the fight-or-flight response.

Cortisol is also involved in the immune response. When levels of cortisol are chronically high, it can cause a loss of lymphatic tissue needed to combat inflammation. Cortisol and progesterone convert protein to glucose, decreasing lean body mass. Further effects of the catabolic hormones include an increase in salts, as sodium, which increases blood pressure and blood glucose and provokes type 2 diabetes and hypertension. The results are a decrease in strength, a decrease in metabolism, and a decrease in attractiveness.

Again, keep in mind that with aging, the anabolic hormones plummet; the catabolic hormones escalate. This hormonal imbalance results in a decrease of lean body mass and an increase in fat. In fact, between the ages of twenty to seventy, you can count on a 45 percent decrease in muscle mass and 60 percent decrease in strength—unless you are working daily to counter that decrease using the brawn-building strategies in this step. As we get older, we must work harder. I tell my patients the work of the retired is to leave the office and get to the gym.

The Limbic–Autonomic Nervous System Balance

Like the anabolic and catabolic systems, the sympathetic and parasympathetic nervous systems are controlled by the limbic or emotional part of the brain. The sympathetic system or "alarm system" is triggered by adrenaline and noradrenaline to get the body ready for fight or flight. The release of adrenaline is like sending a thousand telegrams to different parts of your body at once. These telegrams prepare your body to deal with the stress, whether positive or negative. During the alarm reaction stage, here's what happens in the body:

Blood vessels: The blood vessels in the outer body squeeze shut to force blood to move into the vital organs such as the liver, kidneys, brain, and major muscles.

Blood pressure: As blood flows into the heart, blood pressure soars to circulate blood to the same major organs.

Heart rate: The heart rate quickens, trying to balance the higher blood pressure.

Fuel release: Stored carbohydrate in the liver is converted to glucose (blood sugar) to give a fuel source for the organs that need it, particularly the nervous system and the brain.

Digestion: This process is interrupted.

Frequency of urination: The bladder contracts; urine is released quickly, so accidents can happen.

Lungs: Because of rapid breathing, the exchange of oxygen and carbon dioxide is improved. More oxygen is on hand for the faster metabolic rate, resulting from the stress reaction.

Eyes: The pupils of the eyes dilate to adjust to the dim light (preparing for fight or flight).

Muscle tone: The muscle tone is affected so that trembling is a normal reaction to fear and stress.

When the alarm stage repeatedly occurs, stress manifests itself in disrupting the body. Once the security alarm stops, your body calms down and returns to its former state. The body does this via the parasympathetic nervous system, or the "relaxation system." Whereas the catabolic hormones stimulate the sympathetic nervous system, excreting adrenaline and keeping your body in a continual alarm state, the parasympathetic nervous system makes you calm, decreases your heart rate, and increases feelings of relaxation.

YOUR HORMONAL SYMPHONY IS THROWN OFF KEY

Now, take a seat. The reality of normal aging—that is, aging without the intervention of the 4-Step Anti-Alzheimer's Prescription—is not pleasant, for the good usually gets bad, and the bad sometimes gets worse. Here's what can happen to throw your Hormonal Symphony into disarray:

The Anabolic–Catabolic Balance Reverses

Poor vision (called presbyopia) that happens around age forty is a precursor to the hormonal reversal that hits the human body and brain at midlife. With the anabolic–catabolic balance reversal, you experience weight gain, mood changes, and decreased mental agility. You might notice the following physical changes and symptoms:

- Decreased energy level
- Increased muscle weakness
- Decreased lung capacity
- Increased muscle wasting
- Weight gain around the abdomen

You may also notice the following cognitive changes:

- Difficulty multitasking
- Organizational skills waning
- Forgetting names and telephone numbers
- Decrease in learning ability

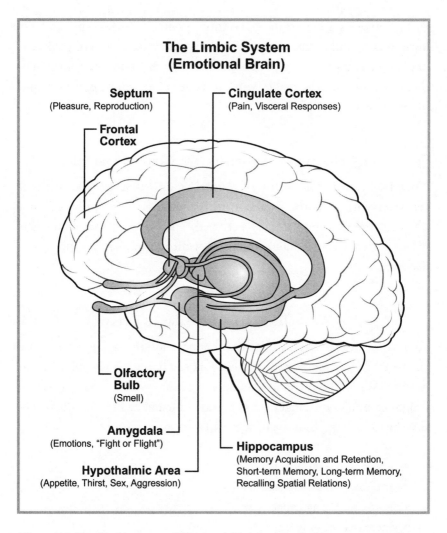

Figure 5.2 The Limbic System (Emotional Brain)—The limbic system controls emotions and emotional responses, mood, and pain and pleasure sensations. It influences the autonomic nervous system, including the vagal and sympathetic nervous systems.

The Limbic–Autonomic System Becomes Less Responsive

Also around age forty, the limbic-autonomic, or emotional, system becomes less responsive. When this happens, you might notice the following signs and symptoms:

- Sleep disturbance
- Overreaction to stress (anxiety and depression)
- Decreased ability to retrieve information

The Cardiac–Brain Balance is Disturbed

When the anabolic-catabolic balance reverses, your heart and vascular system become less responsive, and you may notice the resulting signs and symptoms:

- Increased fatigue with exercise
- Dizziness when standing or bending over
- Hypertension

When your hormones begin to change at midlife, you must resolve to correct the imbalance.

Warning: Only the Brave Read This!

Let's get this straight! According to the latest findings, our bodies lose one-half of the lean muscle mass by age eighty-five. The Baltimore Longitudinal Study of Aging, along with several other reputable studies, shows that lean muscle mass in the body decreases by 10 percent from age thirty to fifty; 25 percent from fifty to seventy; down to 40 percent by eighty years of age; and a giant 50 percent decrease at age eighty-five.

At the same time, the increased fat deposits and connective tissue in the muscles makes them flabby and rigid. And the metabolic rate (something you'd like to stay high so you can burn calories and stay slim) decreases steadily every ten years after age forty-five with an increase in weight if we continue to eat the same diet.

This can be reversed! *By building brawn, you build your brain. By staying strong, you can stay smart and safe.* This is not an aphorism but a substantiated medical fact that you can prove yourself . . . starting today.

We Are Our Worst Enemy

It is more than just a change and imbalance in hormones and neurotransmitters. Along with all these negative changes, we also become our own worst enemy. Take activity level, as an example. Studies show that as we age, our general activity level—meaning our engagement in exercise and physical activities—decreases with the consequent "deconditioning" of our body. In addition, joint injuries to lower extremities and shoulders are six times the incidence from fifty to seventy years old despite the lower activity level. This means you don't have to do much physically to hurt yourself. Also—and this is very important—the social and recreational activities, often a part of exercise that are necessary to decrease stress, diminish *by 50 percent* during this period. (Read more on the importance of having a strong social network in chapter 7.)

BUILDING BRAWN TO BUILD BRAIN: REVERSING THE TREND

So is this deterioration of the body predestined, your fate? Or is there something you can do to reverse these trends? The fact is, exercise makes you look younger and makes you smarter. Accumulating evidence supports what you're about to learn: *Moderate levels of physical activity help to maintain cognitive function during aging.* A recent review of fifteen observational studies that addressed the impact of lifestyle with cognition or Alzheimer's made the following observations:

- An association between physical activity and Alzheimer's disease was reported in six of nine studies.
- A higher risk of cognitive decline or lower cognitive performance was observed in six of seven studies of those who are physically inactive.
- Physical activity, including both aerobic and anaerobic exercise and daily activity, was related to increased cognition in six of seven studies.

There are so many more findings supporting the same, that physical exercise is the best method we have right now to stop and reverse the deterioration of the body—and the mind—with aging. In other words: Building brawn will help rebuild an aging brain.

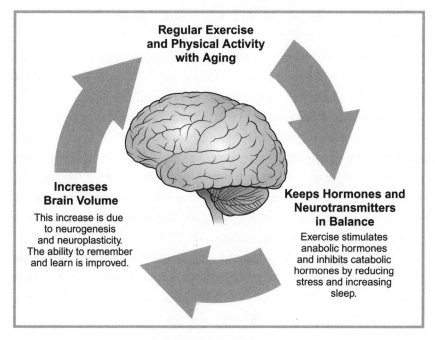

Regular Exercise and Physical Activity with Aging

Increases Brain Volume
This increase is due to neurogenesis and neuroplasticity. The ability to remember and learn is improved.

Keeps Hormones and Neurotransmitters in Balance
Exercise stimulates anabolic hormones and inhibits catabolic hormones by reducing stress and increasing sleep.

Figure 5.3 Building Brawn Will Rebuild an Aging Brain

Understanding the Program:
Building Brawn to Boost the Brain

As a specialist in rehabilitation, I have learned that no matter how healthy you might be, everyone has abilities and disabilities. Medicine is an art, and part of that art is prescribing an exercise program that takes the individual patient into account and not expecting the patient to take into account the "standard" exercise program. *The Anti-Alzheimer's Prescription* differs from other books because it takes into account two vital characteristics of baby boomers: the natural disabilities that we earn with age, and the time deficiency that interferes with our plans to exercise.

Three years ago, Candace, age sixty, came to my office and placed on my desk a book entitled *Exercises for You to Get in Shape*. Candace was a

CPA, still at the top of her career game, yet realized that she needed to get moving to avoid pain and stiffness. At age fifty, she'd had a total knee replacement and now was developing pain in her hips.

"What exercises would you recommend from this book?" she asked.

A quick perusal of the highly illustrated guide made me realize the problem with most exercise manuals and aids. They are not made for anyone older than forty and most of these books use models in their early twenties to demonstrate the movements. The current exercise manuals are not made for baby boomers, especially those who have increasing difficulty with joint pain and problems. Also, it does not take into account the time deficiency that most baby boomers have. That is, despite the best intentions, finding an hour to exercise several times a week is often impossible, especially for those boomers who maintain their own business or work long hours each day.

That's why I was determined to create a *3-step method of exercise* that would be easy to remember, convenient, and conducive for most middle-aged and older adults. As I told Candace, you can do these three steps most anywhere. They include:

1. Stepping with a pedometer (or other moderate intensity exercise);
2. Strengthening with isometric exercises; and
3. Stretching.

The three steps comprise each of the major components of fitness—cardiovascular conditioning, muscular strength, and flexibility—yet, taken individually, provide direct functional benefits as well.

Let's Get Started

Secrets to Successful Exercise

1. Make exercise fun
2. Make exercise safe for you
3. Exercise correctly to prevent injury

At USC we found that making exercise fun, a social event, or listening to music increases the likelihood of making it a lifestyle change by twofold. Music and social interaction gives exercise a double boost, as you'll see in Chapter 7.

Make it safe by being aware of your risk factors. Obesity, large breasts, and joint problems must be considered. See your doctor and get a referral to a physical therapist or qualified personal trainer. Do it correctly to prevent overuse injury and to maximize your effort to get the results you need.

Warm-Up

Prior to an exercise session, you should complete a warm-up (five minutes) based on the activity of which you are preparing to do. Warming up is different from stretching. The goal of warming up is to increase circulation in the body and to begin moving the joints in range of motion. For example, prior to stepping or walking, begin at a slow walk, gradually increasing the rate up to your normal walking level. For biking, start riding at a slow rate and gradually increase your speed. To jog, begin at a slow walk, gradually increasing the rate up to your jogging pace. Prior to isometrics, perform arm swings, leg circles, forward and backward bends at the waist, and simulated motions of the exercises you will shortly be performing.

Exercise and Dieting

Losing weight just by decreasing caloric intake causes you to lose muscle and fat. Dieting and exercise causes you to lose fat and gain muscle. When you just diet and then gain it back and then lose weight again, you just gain more fat each time and lose more muscle. You must exercise each time when on a weight-loss program.

1. Stepping for Cardiovascular Conditioning

The first fitness *S* is for stepping and references cardiovascular conditioning or endurance exercises (aerobics). Endurance exercises are crucial for reducing the chance of Alzheimer's disease, as well as heart disease, stroke, diabetes, obesity, and depression. The general rule is to complete a stepping or aerobic activity you enjoy for at least thirty minutes at a time, three to five times a week. The level of intensity should be such that your heart rate rises to at least 60 percent of your age-predicted maximum (220 minus your age times .6) but not over 85 percent of your age-predicted maximum (220 minus your age times .85). This heart rate range is called the "Training Zone." It helps determine a safe level of intensity to perform cardiovascular training activities while ensuring a physiologic benefit as well.

To minimize the risk of repetitive motion injuries, vary the aerobic activity. For example, you may choose to do stepping with your pedometer three days a week and swim two days. Another way to vary the program is to use different methods that don't tax the same joints and muscles. A routine I use if a gym is available is ten minutes on the bike, ten minutes on the treadmill, and ten minutes on the elliptical. For those without this access, stepping is the best way to get in the exercise you need.

Stepping with a pedometer. With stepping, you attach a reliable pedometer to your pants, belt, or skirt when you first get dressed in the morning, and then attempt to reach a goal, such as walking *10,000 steps* throughout the day. (An inexpensive pedometer can be purchased at most sporting goods stores for under $30.) Here's how pedometer walking or stepping works for conditioning and weight loss:

- 2,000 steps equals one mile.
- One mile burns about 100 to 200 calories.
- 3,500 calories burned equals one pound lost.
- 10,000 steps a day burns 500 to 1,000 calories per day.
- Walk 10,000 steps five times a week to lose one pound a week or fifty-two pounds in a year!

If you have back, knee, or hip pain and find it difficult to step or walk for exercise, that's not a problem. You can opt for at least thirty minutes most days of riding a stationary bike, swimming, using an arm windmill bike, or rowing—all of these exercises will increase endurance and overall cardiovascular fitness. (See www.anti-alzheimers.com for more cardiovascular exercise suggestions.)

Stepping can be done anywhere, anytime. How do you get 8,000 to 10,000 steps in each day, especially if you have a sedentary career? Here's what I suggest to my patients:

1. Park at the end of the parking lot when you shop for groceries.
2. If you use public transportation, get off the bus or train two stops before your destination.
3. Take the stairs throughout your day when at all possible.
4. Take a step break every hour. Some patients get up from their desks and walk briskly in place to accumulate 1,000 to 1,500 steps each hour.
5. Take dancing lessons and spend evenings waltzing or doing the cha-cha, the samba, or swing dancing.

What if you don't like stepping? Again, moderate exercise can include the normal daily activities you must do, such as house cleaning, mowing the lawn, gardening, washing windows, or washing the car—as long as you do the activity at an up-tempo pace, moving briskly. Moderate exercise also includes recreational activities you enjoy doing, such as tennis, baseball, volleyball, golf, swimming, rowing, and jogging.

OTHER MODERATE INTENSITY DAILY AND RECREATIONAL ACTIVITIES

Aerobics (high or low impact)	Biking (both outdoors and
Badminton	indoors)
Baseball	Bowling
Basketball	Calisthenics

Carpentry (moderate pace)	Rowing
Cleaning (heavy)	Running
Dancing	Sexual activity
Dusting (moderate pace)	Skiing
Golfing (pulling clubs)	Soccer
Gymnastics	Softball
Handball	Stair climbing
High-impact aerobics	Stationary cycling
Hiking	Swimming
In-line skating	Tae kwon do
Jogging	Tai chi
Jumping rope	Tennis
Karate	Volleyball
Kickboxing	Walking
Mall walking	Washing car by hand
Mowing lawn	Water exercises
Raking lawn	Yoga (Hatha)
Roller-skating	

2. Strengthening Muscles with Isometrics

The second of the three fitness *S*s is strengthening. For this program, I'd like you to use isometrics as a form of strengthening exercises. Isometric strengthening exercises can stimulate growth factors that stimulate the brain. They require no additional equipment and you can do them any time and any place, including sitting in your car, waiting at a restaurant, or standing in line at the store. Done properly, isometrics can increase muscle strength and functional capacity of the muscles involved in the training.

How-to: Using walls, the floor, or your opposite hand for resistance, simply push (or pull) in a direction at a near maximal effort and hold for a count of 10. Repeat the maneuver five times for each direction prior to choosing a different direction to work. When you have completed all directions to a count of ten five times, go back and repeat the circuit again, thereby completing the workout. Each isometric movement, therefore, should be done at ten-second holds for two sets of five repetitions. See Figure 5.4.

Use caution: To avoid injury, warm up by moving your extremities through full range of motion five times. When first starting isometrics, use 50 percent of maximum effort. After two weeks of doing isometrics, increase the effort to 75 percent. At four weeks, use 100 percent of effort with isometrics. If there is pain or discomfort, stop the exercise completely. If there is still some discomfort upon trying the isometric again, decrease effort to the initial stage (50 percent of maximum).

Unlike stepping (aerobic exercise), which is low resistance and rhythmic, and uses multiple muscle groups at once, strengthening with isometric exercises uses some form of tension or resistance on individual muscle groups. Strengthening exercise is primarily anaerobic (meaning "without oxygen") and is thought by many as best for your brain. The isometric exercises increase muscle mass, reduce blood pressure, and change fat into muscle. Strengthening exercise also increases your metabolic rate and decreases the need for insulin.

Isometric strengthening exercises help to balance hormones. Isometric strengthening exercises help to stimulate the anabolic hormones (thyroid, testosterone, estrogen, and HGH) and to decrease catabolic hormones such as cortisol. When we turn off the adrenal glands that produce cortisol and decrease insulin, we're also decreasing the chance of hypertension, diabetes, and Alzheimer's disease. Strengthening with isometric exercises directly stimulates the brain to produce more growth hormone, testosterone, estrogen, and thyroid hormone—all anabolic hormones or "the good guys"—to protect the brain.

With each isometric exercise, follow these instructions:

1. Hold the exercise position for *ten seconds.*
2. Rest for *five seconds.*
3. Do the exercise again and hold for *ten seconds.*
4. Rest for five seconds.
5. Continue with the exercise until five reps are completed.
6. Move to the next exercise and follow the same time pattern.
7. A total of three sets can be done once you are in good condition.
8. Please read caution advised on page 128 before beginning.

Chest Press

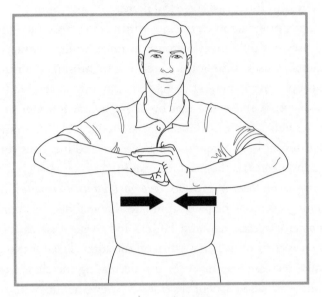

Figure 5.4 Isometric Chest Press

Press your hands together, keeping them at chest height, about six to eight inches out in front of you. Hold this position for ten seconds, and then take a five-second rest. Hold again for ten seconds, and take another five-second rest. Do five repetitions.

Biceps Curl

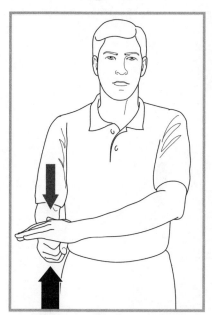

Figure 5.5 Isometric Biceps Curl

Triceps Press

Figure 5.6 Isometric Triceps Press

With your right elbow at your side and palm up, use your left hand to provide resistance. Try to move the right forearm up, resisting the movement with your left hand. Hold for ten seconds; repeat five times. Now change arms and follow the instructions for your left arm resisting movement.

With your elbow at your side, palm down, use the other hand to provide resistance, as you try to move the forearm down, resisting the movement with the other hand. Hold for ten seconds; do five repetitions and then reverse arms.

Leg Extension

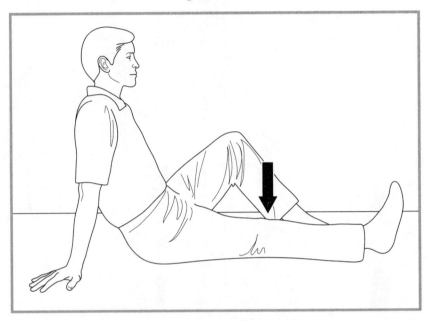

Figure 5.7 Isometric Leg Extension

Seated or standing, tighten the thigh muscle, forcing the knee straight to strengthen quadriceps. Do ten-second holds; do five repetitions and then reverse legs. You can do this seated in a chair with your heels against the floor and buttocks on the chair.

Gluteals

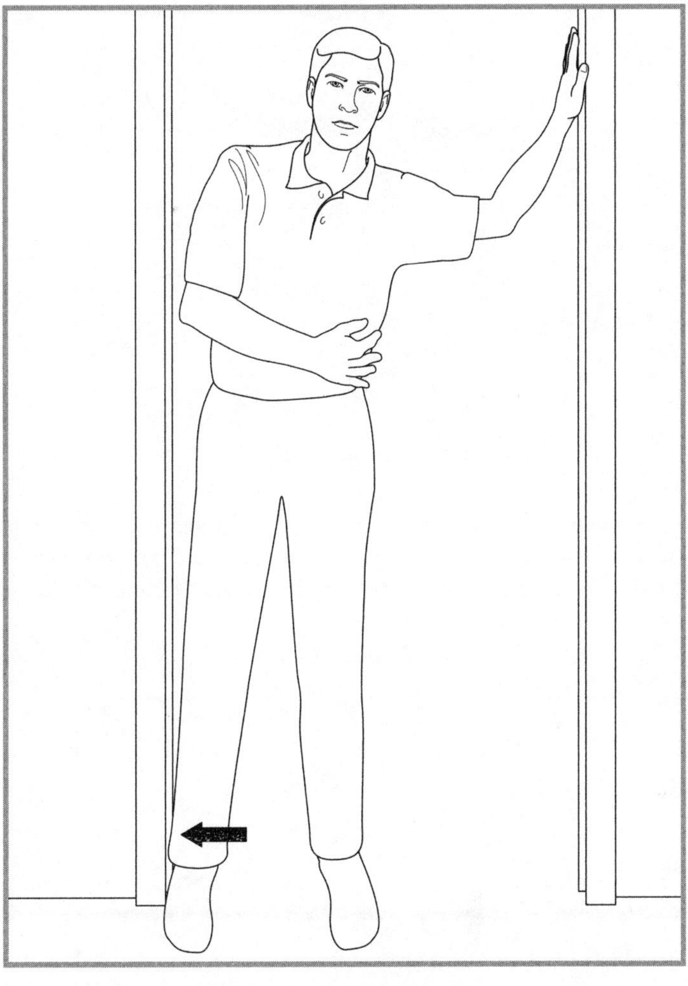

Figure 5.8 Isometric Gluteals

While standing, squeeze your buttock muscles together, trying to rotate your thighs away from each other. Do ten-second holds; five repetitions. This is excellent for preventing low back pain that happens when sitting at a desk. Keep your hips forward with a mild arch to your back.

Abdominals

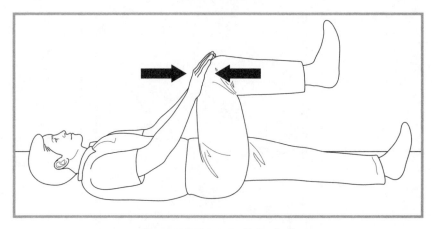

Figure 5.9 Isometric Abdominals

While lying on your back, put your arms on your thighs just below your knees. Try to bend forward into your arms, while resisting or pushing back with the arms. Contract your abdominal muscles as the focus. Do ten-second holds; do five times.

Lower Back

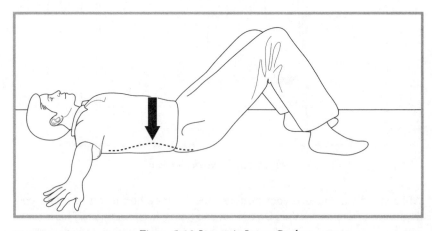

Figure 5.10 Isometric Lower Back

Lying flat on your back on the floor or rug with your legs bent, flatten the small of the back into the floor. Hold this for ten seconds while squeezing your buttocks together. Do five repetitions.

Remember, a muscular body is not because you have made more muscle cells, it is because you have made each muscle cell larger. The same is true about fat. You do not gain more fat cells; you just make them larger.

3. Stretching to Increase Flexibility

The last *S* in the program is stretching. The long static holds typically used to stretch your muscles are best done after you have worked out—*not before a workout.* After a workout, you elongate the recently worked muscles with long static holds, serving as a cooling-down period. Holding the stretch for a period of twenty to sixty seconds (longer is fine as it typically "feels so good") without bouncing or jerking is correct. Shorter holds (ten-second holds, for example), as commonly referenced in magazines and other fitness books, are too short and *not effective.* You will need to hold the position while the muscles are on stretch at least twenty seconds several times, ideally up to sixty seconds two times, for a benefit.

Chest and Shoulder/Waist Stretch

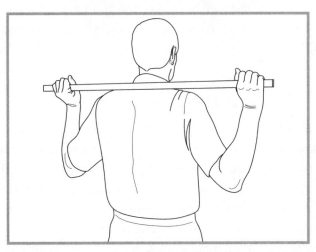

Figure 5.11 Chest and Midback Stretch and
Figure 5.12 Shoulder/Waist Stretch

Place your arms behind your head, stretching your arms back, and then twist left twenty seconds, then right twenty seconds. Rest for ten seconds and repeat three times. You can do more by bending forward or touching your toes.

Leg and Hamstring Stretch

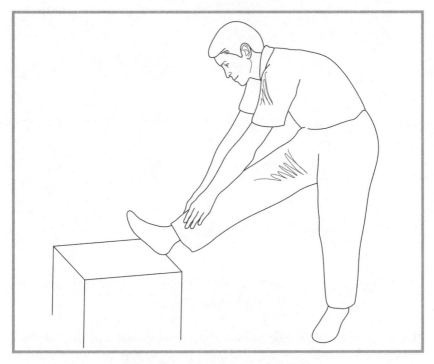

Figure 5.13 Leg and Hamstring Stretch

Stand upright and put one foot on a bench (or step). Slowly bend forward at the hip while keeping your back straight. Alternate, putting the other foot on the bench and repeat the exercise. This is important for decreasing lower back strain.

Back Stretch

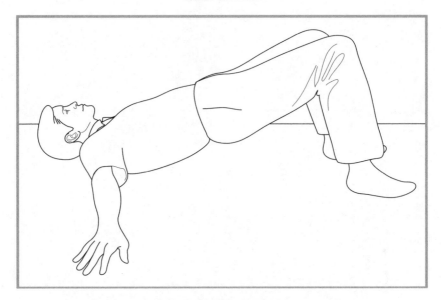

Figure 5.14 Back Stretch

This exercise will help keep your posture straight and alleviate stress on the back and hips. Lie on your back with your knees bent and feet flat on the floor and hip-width apart. While contracting your abdominal muscles, press your lower back against the floor. Then bring your knee to your chest for twenty seconds.

Alternate Circuit Training for Those in a Time Crunch

Circuit training is perfect if you are time-deficient yet want to increase brawn. It's also the best exercise for maximizing calorie burning. Circuit training is both aerobic and anaerobic as it combines cardiovascular conditioning and weight training in the same thirty-minute session. Circuit training includes resistance training as part of the cardiovascular conditioning and should be done *three times a week* (every other day) to be most effective. The following circuit-training activities should be completed in the given order in a consecutive manner, moving quickly from one to the other without unnecessary rest. (If you are in great shape, double this for a sixty-minute circuit-training program.) Circuit training is proven to best control blood sugar in diabetics, as reported in a 2008 Diabetes Association report.

ACTIVITY	WORK LOAD
Stationary Bike	five minutes
Overhead Press, Leg Press, Knee Extension, and Rowing	one set each, up to fifteen reps
Treadmill	five minutes
Hamstring Curls, Upright Rows, Chest Press, and Trunk Rotation	one set each, up to fifteen reps
Stationary Bike	five minutes
Bicep Curls, Leg Press, Trunk Rotation, and Trunk Extension	one set each, up to fifteen reps
Upper Body Ergometer (Arm Bike)	five minutes
Leg Press, Chest Press, Rowing, and Lateral Pull Downs	one set each, up to fifteen reps

Avoid resting between sets and perform the movements in a steady fashion as opposed to pausing at the end of each movement to "stop" the weight from moving. Be sure to move your arms or legs through the full range of motion. When you can perform each set smoothly fifteen times, you may add weight. Do not add more weight than you can support for at least twelve reps. You can substitute a super-slow technique and reduce the repetitions to five. Remember to warm up using arm, leg, and torso movements similar to the exercise you're preparing to start.

To minimize injury and maximize strengthening, thereby increasing muscle with a subsequent increase in metabolism, thyroid stimulation, growth hormone, testosterone, and estrogen, you can use isometrics or do super-slow repetitions to substitute for the regular weight-lifting exercises above.

Super-Slow Strengthening Method

With super-slow strengthening, you use a minimum amount of weight. Each repetition is performed by slowly contracting the weight over a ten-second period through full range, and then relaxing or going in the opposite direction for ten seconds. Do five repetitions of each exercise. I sometimes advise my patients to do one set of isometrics normally and then do another set of isometrics super-slow. Go to www.anti-alzheimers .com for further explanation and a video presentation of the exercise program.

CALORIES BURNED PER ACTIVITY

Activity	30 minutes	60 minutes
Biking, 12 mph	300	600
Circuit Weight Training	400	800
Dance Aerobics	350	700
Golf, Walking	175	350
Racquetball	230	460
Running 12-minute miles	300	600
Swimming freestyle 35 yards/minute	275	550

Activity	30 minutes	60 minutes
Tennis, Singles	230	460
Tennis, Doubles	100	200
Walking, 3 mph, flat	120	240
Walking, 4 mph, flat	150	300
Water Aerobics	150	300

PUTTING THE PROGRAM INTO ACTION:
BUILD BRAWN TO BOOST THE BRAIN

Before beginning step 2, it's important to assess your current health by making an appointment for a physical examination with your health care provider. This medical examination can help you to avoid injury as you start to exercise to build brawn.

1. Assess Your Health

If you are under thirty-five and in good health, there is probably no need for physician involvement with a moderate exercise program. But if you've been inactive, have a history of heart disease, diabetes, or high blood pressure, have a family history of heart disease before fifty, smoke cigarettes, or have a high cholesterol count (total cholesterol greater than 220 mg/dL or HDL cholesterol less than 40 mg/dL), it is important to check with your physician first. In addition, if you answer yes to any of the following questions, the American College of Sports Medicine (ACSM) recommends that you see your physician or health care provider before you begin an exercise program. Consider these important questions:

1. Are you diabetic?
2. Have you ever had chest pain during physical activity?
3. Do you have any joint or bone problems?
4. Do you have asthma or other breathing problems?

5. Do you have hypertension (high blood pressure)?
6. Do you have a heart condition?
7. Do you take any heart or blood pressure medications?
8. Do you ever feel dizzy?
9. Do you have any other medical condition that could interfere with exercise?

Once you have addressed the ten questions and have no health problems, go ahead with step 2. If you do have concerns about your health or if you just want to make sure you are in tip-top shape, go ahead and ask your doctor for a physical evaluation. (Check on page 307 in the Appendix for Strength Test Evaluations you can do without an exercise physiologist.)

2. Take the VO2 MAX Step Test

I'd also like you to take the VO2 MAX Step Test in order to understand how physically fit you are (or aren't!). Although this test is most accurate when performed by your physician, you might try doing it at home. If you have questions on how to perform the test or about the test results, talk to your doctor or a physical therapist.

Take Your Resting Heart Rate. Using a clock or watch with a second hand, take your pulse by counting the beats for ten seconds, then multiplying the result by 6 (12 pulse beats in 10 seconds equals 72 beats per minute: $6 \times 12 = 72$).

Work Out for Three Minutes. Standing at a stairway with a safety bar, step up onto the first step with your foot next to the safety bar. Bring the opposite foot up onto the step. Now step down with the initial foot followed by the opposite foot. Aim to complete this combination of stepping in five seconds, and then continue to step up and step down consecutively for three minutes. Rest for thirty seconds and take your pulse again.

Here's What the Results Mean. If your heartbeat is less than ten beats above your resting pulse (your initial pulse recording), you're at a good fitness level. If your pulse is more than ten beats above your resting rate, you're in average condition. However, if your pulse rate is fifteen beats or more above the initial rate, this indicates fair to poor fitness.

If you have a good fitness level, as determined by the VO2 MAX Step Test, and your physician has given you the okay to exercise, go ahead and start the prescribed exercise regimen given in this chapter. If you're in average shape, as determined by the VO2 MAX Step Test, you should begin at the lowest range of working out and gradually increase to maximum levels. Finally, if the VO2 MAX Step Test indicated that you're in fair to poor fitness condition, it's probably wise to start with an exercise program prescribed by your doctor or begin with the stretching exercises in this step and follow this with the aerobic and then strengthening regimens, as your level of fitness increases.

3. Follow the Anti-Alzheimer's Exercise Plan: The 3 Ss

3 DAYS EACH WEEK (OF MORE IF YOU CAN)	
Time	Type of Exercise or Activity
3–5 minutes	Warm Up
30 minutes	Stepping or Other Moderate Aerobic Exercise
3–5 minutes	Cool Down with Stretches

4. Keep an Exercise Journal

Keeping a regular exercise journal is important as you start the three Ss. Purchase a calendar, and record the amount and type of exercise and activity performed each day, as well as the physical response you might notice (such as difficulty breathing, sore muscles, painful joints). This data can help you and your doctor design a specialized program that has optimum benefit without causing further problems.

5. Be Comfortable

Be sure to try on your shoes with the type of socks that you will be wearing to exercise, and make sure your toes have plenty of room in the toe box. Walk around in the store for a few minutes to see if the shoes rub or are uncomfortable in any way.

6. Start Slowly

Start slowly. Even walking 1,000 to 2,000 steps is okay, if you can do this without injury. As you continue to walk steps, add another 200–400 steps each week. Soon you will be doing the Anti-Alzheimer's stepping workout without struggle or pain.

7. Try Water Exercise

If you find that exercising on land increases joint or muscle pain, check out water aerobics and stretching programs at your Y or fitness center. Make sure you exercise in a warm water pool (about 83 to 88 degrees Fahrenheit), to help warm muscles and joints and decrease pain and stiffness.

KIDS & ALZHEIMER'S PREVENTION

Make exercise a "family habit" for your kids and grandkids. According to Dr. William Sears, renowned pediatrician and author of *Dr. Sears' LEAN Kids* (New American Library, 2003), the easiest habit you can encourage with children is exercising regularly. Make time to walk with your kids. Play ball with them—baseball, basketball, soccer—so they can see that parents and grandparents have fun when they exercise. Another habit to control with your children is television watching. The average child watches seventeen hours per week. If you add computer game and cell phone usage, this is more than doubled. This is not exercise and should be carefully regulated to prevent obesity. When I tossed my first ball to my son and he tossed it back to me, it was more than a ball being shared. We were sharing a love of physical exercise that continues to bond us today. I still remember my dad playing catch with me in Brooklyn at Ebbets Field and in the Bronx at Yankee Stadium. Just writing down this memory allows me to smell the popcorn, hear the organ, and see the the bright, neatly cut grass of the infield as if it were yesterday.

6

STEP 3: BRAIN BOOSTERS

DAILY NEUROBICS TO BUILD A BIG BRAIN'S RESERVE

As the warm, salty sea breeze blew off Long Island Sound amid the scattering seagulls, I walked alone on the pristine sandy shoreline. Suddenly, I was startled as a childhood friend gently touched my shoulder and then touched my cheek and brushed her lips against mine.

I was just eleven; she was twelve. It was my first kiss. Do you remember yours? That same memory, including the scents, the taste of the salty air, the sounds of the seagulls, and the emotional excitement, still vividly stir when evoked by a similar stimulus by the seashore. It is as if the actual event that took place decades ago found a special place inside my brain and continues to live with me daily. This is the magic of memory, the part of our cognitive process that also includes intelligence, language, attention, and executive function.

To understand how to prevent Alzheimer's, you must understand how we remember and retrieve those memories. As discussed, some of the first signs of Alzheimer's are problems with short-term memory and memory retrieval, both of which are essential to learning new material.

HOW YOUR BRAIN CREATES MEMORIES

Right now, I'm going to contradict what you may have learned in elementary science class. In the brain, we actually have six senses, *not just five*. These senses include smell, sound, sight, touch, taste, and perhaps the strongest sense of all: emotion. Each of the senses is stored in different compartments of the brain, and each is vital to the formation of memories. The stronger the sensory stimulus, especially an emotionally laden one, and the more senses employed, the better the encoding of that memory. The best way to teach children is to tell them, show them, and then ask them to do it. Creating memory reserves is no different. It is not enough to hear music; we must sing the song, play the instrument, and share the music in order to keep it in our memory.

I often explain memory as a complex of rooms that, when unlocked, reveal the extraordinary gift inside. Opening one room allows entry into all of the others, re-creating your memory as if it happened that very moment in time. Our nostalgic re-creation of specific memories and their interaction is possible by way of dendrites—communication wires that are now believed to store the memories—along with their synapses and neurotransmitters. As an example, a small sensory stimulus such as a smell (salt air) or the sight or sound of a seagull can stir a memory cell to open even unconsciously. When this happens, the brain re-creates a very vivid and specific memory that was chiseled into your dendrite bulbs many years before—whether of a relaxing vacation, a parent's embrace, or that first kiss.

Memory is modulated by a complex number of neurons, dendrites, synapses, and neurotransmitters such as glutamate and acetylcholine. In fact, the real tough work of your brain goes on in individual cells—about 100 billion nerve cells or neurons with branches that connect at more than 100 trillion points. We call this thick, branching network a "neuron forest." Whereas signals traveling through your neuron forest form the basis of memories, thoughts, and feelings, neurons are the main type of cells destroyed by Alzheimer's disease.

In anticipation of the future, specific neurons and their connecting wires, which are maintained even as we age, neatly separate our chronological memories, both past and present. However, these well-defined and separate wires and storage stages can also be disturbed, which is what happens with Alzheimer's when the destructive beta-amyloid proteins attack the pathway dendrites, dendritic bulb storage areas, axons, and nerve cells. These wires, synapses, and neurotransmitters remain as long as they are used. The more they are activated, the stronger the connection and the larger the pathway—thus, the more easily these memories can be retrieved. If the dendrites and bulbs diminish in number, it makes it difficult to access the stored information (as happens in Alzheimer's).

As the plaques and tangles expand, the distinct roads or wires break down, as do the special walls or compartments that separate the past and present. As my father and millions of Alzheimer's patients experienced, the past and present all seemed as one place in time. For instance, when my father was in stage 3 Alzheimer's, he could no longer remember that his father had died. He no longer knew where he lived, the names of his children, and, sadly, he no longer recognized my mother, Rose, as his wife. When the wires break down in the brain, the Alzheimer's patient suffers with frightening illusions, hallucinations, and an inability to understand and reason with the present. That's because the present is dependent upon our remembering both recent and remote past memories.

BRAIN MATURATION

So what happens with aging? Must we suffer with these incomprehensible problems, including loss of memory in older years? While many people think that brain aging starts in midlife or later, neuropsychological testing of healthy young adults shows that mental agility starts its decline when we're fairly young, about age twenty-four. Then these same tests note that every seven years after age forty, the brain's processing speed slows down appreciably.

Well, if this is the case, then most of us baby boomers are in our second or third mental agility step-down. Must this *always* be the case? Maybe not! There is a vast—and important—difference between mental agility and mental capacity. There is no doubt in my mind that what I know now (my overall mental capacity) far outweighs what I knew at twenty-four. Yet my agility, my ability to process and learn, is definitely slower . . . as are my legs!

As proof that our mental capacity may actually increase with age, let's look at how some individuals perform their greatest creative work late in life. Verdi, for example, composed *Othello* at age seventy-three and *Fal-*

staff at seventy-nine. Humboldt wrote the five volumes of *Kosmos* between the ages of seventy-six and eighty-nine. Goethe produced the second part of *Faust* when he was more than seventy years old. Likewise, Galileo and Sherrington continued to make scientific contributions in their eighth decade of life.

So how does this make sense—that mental agility declines with age yet mental capacity increases in many individuals? We do know that high intelligence, well-organized work habits, and sound judgment compensate for any progressive mental deficiencies in the later periods of one's life. Let me give you an example of my close friend, Dr. Marshall Wells.

At ninety-four, Dr. Wells was still seeing patients and relayed the history of a confused and disoriented patient—also a doctor—who was confined to a hospital bed. Without a chart to rely on, Dr. Wells succinctly gave the full past medical history of his younger colleague, complete with the medications Dr. Marks was taking, including name brand, generic name, and the dosages and frequencies. Dr. Wells apologized for any inaccuracies, saying, "At ninety-four, I'm not as quick as I once was."

Recently I attended Dr. Wells's 100th birthday celebration. Standing before a crowd of 600 friends, colleagues, and former patients, he gave a most eloquent speech about his memories of being a missionary physician. He talked in detail about the Bataan Death March in the Philippines, his four years of imprisonment by the Japanese after the bombing of Pearl Harbor, how he opened the first hospital in Thailand, and his extensive missionary work in China over a seventy-year period. To me, and to those who know him, Dr. Wells's life exemplifies the 4-Step Anti-Alzheimer's Prescription.

Now it's important to realize that as we age, not all brain changes are bad. In fact, many changes in the aging brain are necessary and helpful. For example, prior to the age of four, our brain cells are constantly differentiating and maturing. If a child suffers a brain injury during the toddler or preschool years, it can often self-repair. As an example, at age three and a half, our daughter, Kaycee, suffered with encephalitis, which resulted in damage to the brain that left her completely paralyzed on one side. I can see her right now, as if it were yesterday, as the right side of her precious face drooped downward, and she was unable to speak intelligibly. Kaycee could not move her right arm and dragged her right leg when she attempted to walk. This horrible memory is seared into my brain.

But a child's young brain can also self-repair. At sixteen, Kaycee won, for the second time, the National Junior Karate Championship. As for her speech and intellect, she now has a master's in psychology, and communicates very well.

Between four and ten years of age, we have a great facility for learning languages. When we learn language at this age, we are not left with an accent that might give someone a hint of our original heritage.

Then, at age twenty-four, our brain's agility peaks. If you look back over history, you'll see that most scientific and mathematic discoveries, including Einstein's Theory of Relativity, occurred before the person was twenty-four years old.

How the Brain Works—And Why It Sometimes Forgets!

There are three main areas of the brain—the hindbrain, midbrain, and forebrain (see figure 6.1). The hindbrain and midbrain contain structures that process body functions like heart rate, breathing, balance, coordination, muscle movement, and some sensory information. These types of functions are considered "lower" brain functions because they're automatic—that is, we don't have to think about doing them. The forebrain processes both lower brain functions like hunger, thirst, and sensory information, and "higher" brain functions such as memory, logic, conscious thought, vision, hearing, and language. Included in the forebrain is the limbic system, a network of structures that process emotion. I first defined the thinking brain (neocortex) and emotional brain (limbic) in chapter 1, page 9.

The secret to success in the Anti-Alzheimer's Prescription is maintaining the thinking brain's control over the emotional brain. It is like keeping the superego in control of the id. Scientists have found that different areas of

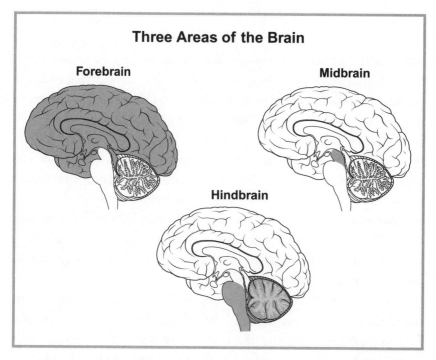

Figure 6.1 Three Areas of the Brain: Forebrain, Midbrain, and Hindbrain

The forebrain includes the neocortex, or thinking brain, and the limbic system, or emotional brain. The midbrain and hindbrain control automatic functions such as breathing, balance, coordination, and heart rate.

the brain have specific functions. For instance, while reading this page you are using an area of your brain termed the "visual cortex." Neurons pick up information from your eyes and then send it to the visual cortex at the back of the brain where the information is processed and sent to other parts of the brain. (Its location is why people might "see stars" when they hit the back of their heads.)

Many of the stereotypical glitches we associate with aging, including being forgetful, misplacing important items, having difficulty learning new concepts, or simply not feeling on top of our game, involve two specific areas of the brain: the cerebral cortex and the hippocampus, two areas of the brain that play a vital role in learning and memory.

The Cerebral Cortex

The cerebral cortex contains an astounding number of nerve cells—about 100 million per square inch. All of our sensory information, whether from taste, touch, or smell, arrives at its own dedicated area of the cerebral cortex. Sight has thirty such specialized areas. It takes a network of many smaller regions of the brain to process all of the information flooding in from the senses. These hundreds of regions are linked together by the brain's equivalent of wires, thin threads called axons and dendrites (each only one-hundredth the thickness of a human hair) that link nerve cells. The brain contains literally hundreds of miles of such wires. This network of pathways between the various cortical regions allows the cortex to be adept at forming associations with the process of creativity, deductions, logic, and conclusions.

Specific areas of the brain that degenerate with aging are the locus coeruleus and substantia nigra. I know, these sound like the name of a Roman legion or the name of an ancient river in Egypt! Actually, the locus coeruleus (meaning "blue place") is in charge of our physiological reaction to stress and panic. It releases norepinephrine, a neurotransmitter associated with alertness, concentration, aggression, and motivation. Degeneration of the locus coeruleus cells can cause a "dull" feeling that can't be cured by a strong cup of coffee. In Alzheimer's disease, loss of these cells causes a person to be apathetic or seem disinterested. Alternately, the substantia nigra is responsible for our ability to control our movements. Degeneration of the substantia nigra cells can lead to a disorder such as Parkinson's disease, which also runs in my family.

The neocortex (your thinking brain) is particularly affected by lack of use and will shrink, or atrophy, if not stimulated. This atrophy, like muscular atrophy, usually occurs after age sixty. As I've stated previously, the neocortex is responsible for many functions, including sensory perception, spatial reasoning, conscious thought, language, and executive function. However, the most dramatic loss to this brain region occurs in the small neurons of the second and fourth layers and the frontal and superior temporal gyrus regions, which can lose up to half of its cells by the time we are in our nineties. When these areas in the neocortex deteriorate, we lose our ability to understand speech, to understand what we see, along with the recognition of familiar objects, faces, and places. This loss, called agnosia, comes from the Greek word *gnosis* meaning "knowledge." A similar occurrence in motor areas of the brain, called apraxia, causes us to lose our normal abilities to dress, groom, and, eventually, walk. (*Praxis* means "movement" in Greek.) In apraxia, despite having the motor ability to do an action, we have forgotten how. The late stages of Alzheimer's disease are marked by apraxia and agnosia; in other words, we lose our ability to do the activities of daily living.

> Agnosia is the loss of knowledge of previously known facts, which include everything from recognizing a pen as an instrument to write with and the face of one's own spouse.

The Hippocampus

Meanwhile, the hippocampus (a small bean-shaped area in the brain) receives the barrage of sensory information from the cortex and starts filing it away. To prevent the information overload that would accompany having to retain the entire influx of information, the hippocampus sifts through and picks out which information to store and which information to discard. The hippocampus's decision to store a memory is believed to hinge on two factors: whether the information has emotional value and whether it relates to something the person already knows. I've already told you about my first boyhood kiss—definitely an emotional moment that is stored as such in my brain.

Most of what we learn and remember relates to the brain's ability to form and retrieve associations. For example, if you pick a deep-red rose while on a stroll in the park and inhale its scent, its image, feel, and smell activate specific areas of the brain. Once that happens, any experience that activates the areas of the brain stimulated by picking and smelling the rose will then trigger all the areas of the brain's regions that have ever been activated by representations of roses. In other words, if on another occasion, someone hands you a deep-red rose, you may suddenly remember the day you took the stroll and picked the first red rose. Or this new rose may remind you of Aunt Harriet's rose garden and how you enjoyed smelling the different roses when you were a child. Then, again, the new rose may bring back memories of a time in a previous marriage when your former spouse sent you wilted roses. In this last instance, the memory may not be so pleasant—but it is still a viable memory embedded in your brain. The brain will almost automatically link sensations, and, in essence, this is our basic learning process. Unknown to most of us, ad agencies and marketing gurus exploit these unconscious hippocampal limbic preferences (*our likes and dislikes*) in their attempt to sell you their products.

The hippocampus also controls the information we don't want to retain—or what we forget. This process is important because, despite our big brain's capacity, it would be overwhelmed if it retained everything we initially learned. Also, if the hippocampus kept all information, it would interfere with our ability to prioritize and organize our thoughts and memories. The hippocampus might be seen as the gatekeeper of all new information. Due to the incredible amount of information it's assessing every second, the hippocampus has a high metabolic need for energy supplied by the brain's blood supply. Any loss of blood pressure, change in nutrients, or even hormonal changes from an increase in cortisol affects the hippocampus and, consequently, learning and memory.

Anatomical Deterioration of the Brain

Let's look at the physiology of the anatomical deterioration of the human brain that occurs between the ages of twenty-four and eighty:

- Brain weight decreases by 15 to 20 percent.
- Blood flow to the brain decreases by 20 percent.
- The number of fibers and nerves decreases by 37 percent.
- Brain volume shrinks 0.5 to 1% every year after age sixty-five.
- Patients with Alzheimer's may lose 3 to 5% of their brain volume per year.

Some areas of the brain may shrink more than others. For example, the frontal lobe (important for mental abilities) shrinks. The neocortex (thinking brain) is progressively depleted in the seventh, eighth, and ninth decades, and there's a loss of cells in the limbic (emotional) brain, especially the hippocampus, where new memories are processed. The hippocampus experiences a linear decrease of more than 25 percent between ages forty-nine and eighty. White matter also decreases. The brain's white matter is made up of myelin, a fatty white substance that helps to protect and improve communication speed between brain cells and dendrites. Changes in white matter are linked with changes in the speed of cognitive processing (memory, action, problem solving, decision making, and attention) or mental agility.

Yet in the midst of losing some function in certain parts of the brain, our personal and emotional skills continue to develop with age. *What we call "wisdom" evolves throughout a lifetime of experiences. It is often said that experience first gives us the consequences, and then it teaches us the lesson. I believe it is in the accumulation of these lessons that we acquire wisdom.*

NEUROBICS: GIVING YOUR BRAIN A WORKOUT

Twenty years ago, I could remember telephone numbers told to me once and I could keep them in my memory for years. Today, I think I am a genius if I can remember in the morning where I left my wallet, cell phone, and keys the previous evening.

How about you? Have you ever forgotten your car keys only to find that you'd put them in the refrigerator when you poured yourself juice earlier that day? Maybe you spent all morning writing a detailed grocery list only to get to the store and realize that you'd forgotten to bring it. A colleague admitted that she often forgets where she puts her reading glasses and invariably searches her office in a panic, since she cannot see the computer screen without them. "Usually a kind employee will gently tell me the glasses have been on top of my head the entire time," she reports.

If you're like most people, you might immediately assume that forgetfulness is a sign of the brain's inevitable decline. Yet, some revealing findings are giving neurologists like me new clues about brain aging—and what we're realizing is *the brain does not have to weaken as we get older.*

It was in 1992 that neurobiologist Evan Snyder, M.D., and his team of Harvard researchers came upon stem cells in the brain. These revolutionary findings contradicted the common belief that brain cells age just like other cells in the body. In addition, the scientists also realized that if brain cells aren't regularly switched on, it can result in atrophy of the dendrites and cells. Dendrites are the branches on nerve cells that directly receive and process information from other nerve cells, forming the basis of memory, a process called long-term potentiation. Thus, the mental decline most people experience with aging may not be from the steady death of the brain's nerve cells. Rather, mental decline is due to the thinning out of the number and complexity of dendrite bulbs. The more we challenge the brain, the more dendrites and bulbs we create and therefore the more memory and connection we make. Thank goodness, we're learning tangible ways to strengthen the brain and prevent its early decline. And that's where accentuating the brain's reserve (neurobics) comes into play.

Simply put, neurobics means giving the brain a real workout—every day of your life. Neurobics is designed to help you maintain a consistent level of mental fitness, strength, and flexibility as you age.

There's accumulating evidence, mainly from longitudinal observational studies, that higher levels of physical and mental activity and social interaction can help maintain or even improve cognitive function during aging. That's because the networks of blood vessels keep oxygen flowing to the billions of brain cells. Branchlike tentacles extend from the ends of those cells, the brain's own communication "wires." A healthy brain—*no matter how old*—can continue to grow new neurons and dendrites and rewire itself. A major source of memory capacity is found in the dendrite bulbs that can increase in size, thus increasing memory storage (your interest-bearing account).

Neurobics Help Maintain Mental Fitness

Neurobics can help you maintain a consistent level of mental fitness, strength, and flexibility as you age. By presenting your brain with nonroutine or unexpected experiences that use all your senses, neurobics create new patterns of neuron activity and improve the brain's ability to form associations. Associations, such as putting a name with a face or a smell with a certain food, are the building blocks of memory and learning. Deliberately creating *new associative patterns* is a central part of neurobics. In the future, along with gymnasiums, we will have *neurasiums*, which will let you be as time efficient as possible. As an example, many of the neurobic exercises can be performed while building brawn, a perfect example of multitasking that accentuates the brain's reserve even further.

Jane, forty-nine, had recently moved from Chicago to Los Angeles after the death of her husband two years prior. Since her husband's death, Jane started forgetting appointments and became increasingly isolated in her apartment, watching TV all day and rarely calling her only daughter. Despite the antidepressants I prescribed for her malaise, Jane continued to have word-finding difficulties. While the neurological examination was normal, Jane's responses were somewhat slow, certainly less than I would expect for her age. I made a tentative diagnosis of mild cognitive impairment. What I believed was that Jane was accelerating her mental decline due to her sedentary, solitary lifestyle and lack of mental stimulation. TV is not stimulation. Jane, though not under apparent stress, was in fact stressed because of a lack of purpose. Sleeplessness was self-prescribed as she watched TV till 3:00 A.M. I gave her a memory profile, which included an immediate mental agility evaluation. Her findings suggested, again, mild cognitive impairment.

I advised a new diet, exercise program, neurobics, and a socialization program. We instituted a neurobics mental exercise plan, which included navigating her apartment with her eyes closed, buying a Nintendo game called Brain Age, and learning Spanish at a nearby community center. Every evening she would take a thirty-minute walk with a neighbor that was divided into three ten-minute segments. At the end of the first ten minutes, they would close their eyes and pay attention to experiencing whatever they heard, smelled, or felt. They then discussed this over the next ten minutes. During one of these sensory accumulating experiences, Jane noted a smell of a sauce that reminded her of her childhood. They pursued that smell and three blocks away they found a German delicatessen that in fact made the same sauerbraten that she had been brought up with.

When I saw the patient three months later, her memory difficulties were over. When I retested her mental agility and short-term memory, she had completely normalized.

New experiences such as navigating an environment using different senses other than sight actually cause nerve cells to produce natural brain nutrients called *neurotrophins*. Neurotrophins make surrounding cells stronger and more resistant to the effects of aging, and can increase the size and complexity of nerve cell dendrites. In doing neurobics daily, you'll help make your brain more agile, flexible, and healthy so it can take on any mental challenge, whether a memory, task, performance, or creative project.

Engagement and Novelty Grow New Synapses

Recent research suggests that engagement in meaningful activities may help maintain cognitive functioning with advancing age. *Engagement can be in activities that require active new learning, analysis, problem solving, or agile thinking*—pursuits like singing, learning to sketch or watercolor, or mentoring children at a nearby elementary school. You can study a foreign language, finish a college degree or advanced degree, take a decorating course, or learn a new musical instrument. I recommend to my patients that they take advantage of the classes at their local community centers as Jane did, do crossword puzzles or sudoku daily, or join neighborhood book clubs—all of which are new experiences in their lives and require them to think on new and different levels.

Engagement can also be making a commitment to new challenges— volunteering at a local hospital, helping with organizations such as Habitat for Humanity, or offering your time and energy to a to a faith-based group to assist in local projects. All of these activities help to keep your brain active and alive, especially if you're considering retirement from your full-time work.

Novelty, meaning freshness and uniqueness, is also important for keeping your brain alert with aging. Novelty means viewing your normal daily routine in a new or different way. Stepping outside of your daily routine, or relying on your senses in new ways as my patient Jane did, can help to keep your brain active, growing, and healthy.

I challenge you to find novel ways to experience your environment by finding a new route to work or to the mall. When riding the train to work, close your eyes and try to distinguish various sounds, smells, and feelings. Rather than sitting there half asleep, consider what you might have been doing a decade or two ago. Let your brain fall into its full creative mode as it challenges itself to think in a different way than normal.

Olfaction Plays a Key Role in Memory

Nowadays, people depend primarily on their senses of vision and hearing to assess their environment, whereas a keen sense of smell was once vital for survival. Farmers could smell a change in weather and doctors previously used smell to diagnose illness. Despite olfaction's diminished role in our daily lives, it plays an important role in memory. Associations based on odors form rapidly and persist for a very long time. The olfactory system is the only sense that has direct connections to the cortex, hippocampus, and other parts of the limbic system. That's why certain aromas like pumpkin pie and chocolate chip cookies trigger an abundance of emotional responses. The sense of smell begins to deteriorate starting at forty, by as much as 50 percent by age seventy, as does one's taste. Serious loss of smell can be the harbinger of Parkinson's and Alzheimer's. A sudden loss of smell, however, can be seen most commonly with respiratory conditions or after serious head trauma. Try this: Close your eyes and smell each food item on your plate and visualize it. Recent studies show this builds your olfactory reserve.

A Strong Social Network Challenges the Brain

Memory is also largely dependent on its emotional context. The hippocampus is more apt to tag information for long-term memory if it has emotional significance. Interactions with other people are an important trigger of emotional responses. In fact, *research shows that people who have a close network of friends and spend time with them tend to live longer and healthier lives and stay mentally sharper.*

As we age, our social circles tend to shrink and so do the areas involved with social interaction. *What we don't use we lose.* An important aspect of neurobics is to find opportunities to interact with others. A revealing study published in the prestigious journal *Lancet* reported that after following 1,200 people in Stockholm, Sweden, for three years, researchers found that those individuals with a *limited social network* had a 60 percent higher risk of developing dementia and Alzheimer's disease. Other findings show that having a good long-term relationship at age fifty correlates with good health at age eighty.

Social interactions themselves have positive effects on overall brain health. Engaging emotions through social interactions is also a key strategy of neurobics. Working out at a gym or with a friend challenges the brain and lowers stress levels, thereby improving one's likelihood of building the brain, a double effect. Double effect is combining one's efforts.

What Happens to an Unfit Brain (Age Thirty and Beyond)

Before you begin the suggested neurobics, I want to explain why it's important to keep your body and brain in shape. Just as you work out to keep your body fit, daily mental crunches are especially important once you're in your thirties and beyond. From then on, the brain is more susceptible to deterioration. Unfortunately, a weakening brain may undergo dramatic changes. The brain can lose up to 20 percent of its weight and blood flow, as well as experience a decrease in its nerve conduction velocity—all of which slow down the thinking process.

Most people who have an unfit brain also have an unfit body. With aging, the amount of oxygen your body is able to use during a jog, walk, or swim can decrease by about 60 percent. So, too, can the number of taste buds in your mouth, not to mention that your metabolism slows down and hand grip weakens. This means you must work harder to produce the same results as when you were young.

Before you close the book and think it's too late to spark what you feel is a dull mind, here's some good news: Studies show that surviving neurons exhibit expanded dendritic trees, suggesting that even aging neurons have the capacity to react to cell loss by growing new synapses. New synapses are important because they are the pathways that create new associations and a stronger mind.

MAKING NEW BRAIN CONNECTIONS

An important part of my brain-boosting neurobics strategy is to help you use your senses to increase the number and range of associations available to you. Because each memory is represented in many different cortical areas, the stronger and richer the network of associations, the more your brain has some protection if it should lose neurons and dendrites. *The larger your brain "safety net," the greater the chances are that you can solve problems or meet challenges.* That's because you have many more pathways at your disposal from which to reach a conclusion.

Let's look at a very common problem associated with aging and with Alzheimer's disease, that is, remembering names. Imagine that you are meeting a new client, your child's (or grandchild's) new teacher, or a new neighbor. Instead of listening to the person's name and shaking their hand, I'd like you to focus on the following:

1. Focus on the feel of their hand. Is it soft, rough, cold, warm, or clammy?
2. Focus on their smell. Is it shower fresh? Are they wearing a scent like musk?
3. Focus on their voice. Is it deep, loud, soft, or shrill?
4. Focus on a unique characteristic about them. Are they quite tall

or short? Do they have a broad smile, a mole on their cheek, unusual eye color?

5. Focus on past memories and associations. Does the person remind you of someone, perhaps a favorite aunt or a dreaded teacher?

You have now tagged someone's name with not just one or two associations, but with at least four *very personal connections*. This is extremely important in maximizing your brain reserve for later in life. Adopting strategies such as this for forming multisensory associations when the brain is still at or near its peak performance—in the forties and fifties—builds a bulwark against some of the brain's inevitable loss of processing power later in life.

BUILDING RESERVES IN YOUR PERSONAL BRAIN BANK

While many think the brain is a fixed bank account that we can draw on until it goes bankrupt—a dwindling, stagnant organ that slowly loses its cells and connections until it stops functioning—this is far from the truth. Your brain's reserve has the ability to create new neural pathways and connections (neuroplasticity) and new cells (neurogenesis) that can be used as a mental savings account, a reserve to be drawn upon in time of need.

The brain's reserve is unlimited because of stem cells and its ability to form dendrites (connections) when stimulated. *If the brain is healthy and constantly stimulated, it can and does continue to grow throughout our entire adult lives.* Comprehensive studies show that the more educated we are and the more we use and challenge our brains by taking classes, traveling to new destinations, learning a new instrument or foreign language, or playing the challenging, brain-boosting video games, the deeper our brain reserve gets. Brain-boosting video games include Nintendo's Big Brain Academy, Brain Age, Mind Fit for PCs, and Mattel's Radica Brain Games.

Unlike the outward signs of bodily aging—the wrinkled skin, graying hair, tooth loss, decreased vision and hearing, and stooped posture—

the first sign of brain aging is not obvious. First, there's the slow decrease in interest such as my father exhibited when he was indifferent to me at the airport and lost his mental agility and wit. *These initial symptoms of brain aging are followed by an inability to learn, adapt, and respond.*

With brain aging, there is also a decrease in recent, or short-term, memory. For my father to forget the baseball game at Dodger Stadium was a clear red flag. After that, changes in long-term or remote memory become clear. *Because memory is the most quantifiable measured factor of cognition, we neurologists use it as the diagnostic tool* much as we use an EKG to determine cardiovascular disease, pulmonary function tests to evaluate lung capacity, or laboratory tests to measure blood lipids and C-reactive protein, a measurement of inflammation.

More often than not, adults don't exploit the brain's rich potential for multisensory associations. Most of us go through our lives engaged in a series of remarkably fixed routines. By doing this, we reduce the number of opportunities for making the new associations that are ideal for mental fitness. Still, the human brain is evolutionary, primed to seek out and respond to novel information. Neural pathways activate when the brain processes an unusual task rather than when it performs a routine one. During a routine task such as folding clothes, putting the dishes away, or stapling papers together, there is no increased activity in the anterior cortex, cerebellum, or frontal cortex. Such passive stimulations of the senses simply don't exercise the brain.

Putting the Program into Action:
Use Daily Neurobics to Build the Brain's Reserve

There is nothing magical about neurobics. Rather, the magic lies in the brain's remarkable ability to convert daily mental activities into its neurobic exercises. Simply by making small changes to your daily habits, you can add mind-building exercises into your daily life. Still, no brain exercise program is going to help you if you aren't motivated and committed. That's why neurobics is recommended as a lifestyle choice to use in combination with steps 1, 2, and 4 of this program.

So often, we assume that elderly adults don't listen because they are hard of hearing or uninterested in the world around them. Instead, their brains may simply not be getting the intentional workout necessary to stay alert. Older adults who think they are stimulating the brain by doing simple reading may need to do a lot more mental activities to build brain reserve, since activities that are done repeatedly become second nature. *Reading without actually thinking and interacting with other parts of the brain does not build much brain reserve.* With each subsequent try, the activity demands less of what's necessary to keep the brain's acetylcholine system tuned up (this is the brain's attention-getter that dominates the circuits used during paying attention and focusing). When you are unable to pay attention, you'll have more trouble with short-term memory. And because of the centrality of attention to neuroplasticity, a brain that fails to pay attention is a brain that cannot tap into the power of neuroplasticity. So rather than doing the ritual games or activities, engage yourself in activities that are new, that require your attention because of their novelty—as I've given you with these neurobics.

MEMORY

The key to memory exercises is repetition. A kitchen timer may be helpful with neurobics, since many of the following activities rely on time and duration. An old wind-up timer is fine, but a digital display timer might be easier to read.

Number Memorization

We have many lists of numbers locked in our memories, from phone numbers to birthdates to addresses. In this *Number Memorization*, I'd like you to continue memorizing lists of numbers. Start small with six numbers. Then, challenge yourself with ten- to twelve-digit number sequences.

Allow one to two minutes to memorize the sequence of numbers, and then test yourself throughout the week to see how long it takes to remember six, then nine, and then twelve numbers and more. (As you master memorization of six numbers, challenge yourself with seven—a phone number—and then nine and twelve.) Here are some examples:

- Your siblings' anniversary dates
- Your siblings' spouses' birthdates
- Birthdates of coworkers
- Your neighbor's phone number
- Your license plate number
- Your neighbor's license plate number
- Credit card numbers
- Your favorite team's batting averages
- Pi to the thirtieth decimal place (3.14159265358979323846 2643383279)

List Memorization

This memorization activity is fun to do with a group such as a gathering with your children and grandchildren, neighbors, or even coworkers. Remember *The Twelve Days of Christmas* song with "four calling birds, three French hens, two turtledoves?" This game follows the same premise as the song.

Have someone in the group start the list with one "something"— such as "one barking dog." The next person in the group has to remember and then repeat "one barking dog" and the following person then adds two items—such as "two feisty cats." The next person then remembers the "one barking dog, two feisty cats" and adds three of "something." See how long you can make the list. I had a patient who could remember and repeat twenty-nine items—even at his appointment a week later—and he was seventy-nine years old! If you are doing the game alone, try memorizing your grocery list—and then repeat it later at the store.

Take the Road Less Traveled

This memorization game is a good one for road trips. If you're traveling to a place you have been to before, and are familiar with the route you usually take, I challenge you to select a different route and see if you can still figure out how to get to your destination.

Another way to play this game is if you take one route to work, use a map Web site to plot a new route. Then *memorize that route.* This superb brain stimulator introduces new environments that increase sensual awareness.

If you're an avid walker, walk a new route without planning your way and see if you are able to make it back to your original location faster. This is even more beneficial, because exercise stimulates the brain and builds brawn!

Brain Dance

Dancing is excellent for the brain and body. Not only are you moving around more and giving your heart a great cardio workout, but you must also memorize specific steps and apply those steps to a certain rhythm and beat in the music.

Ballroom dancing, swing dancing, the foxtrot, and salsa dancing all help to improve your endurance, coordination, and balance—and your brain is in constant motion, as it recalls the proper steps and movements. You can enroll in a class at your local dance studio or go to a video store and rent "how to dance" videos or DVDs. You don't even need a partner; you can learn the steps yourself. If you are physically unable to do something like the foxtrot, try going through the motions in your mind or moving your arms while visualizing the memorized dance moves.

Dictionary Words

This is a perfect early morning neurobics challenge to enjoy with your coffee or tea. Each day, select one word in the dictionary that you are unfamiliar with and write it on a three-by-five index card. Memorize the spelling and definition, and challenge yourself to use the word five times during the course of the day. As you become more proficient with memorization, give yourself a time limit, say sixty seconds, to memorize the spelling and definition of the word. When you've mastered that, try to do this in thirty seconds. Not only will this increase your brain's reserve and memory, but you will be broadening your reach of the English language. To add more challenge to this neurobics, do the same with a new foreign language.

Visualization

Visualization stimulates the brain as you go deep into the wellspring of your memory and imagination. Not only must you recall a memory, but also you must investigate the memory with all your senses: the sights, smells, tastes, sounds, feelings, and emotions.

For you golfers, buy an instructional golf tape. Then in the following days, close your eyes, imagine yourself performing the swing. Remember the tempo, the follow through; imagine the little white sphere exploding out of sight as your swing speed reaches 150 miles per hour. You will see your game get better and your imagination explode. The older we get, the better our imagination can become.

I Spy a Memory

This neurobics exercise is great when traveling on long road trips. Traveling naturally stimulates the brain because of the constantly changing sights, sounds, and environment. While traveling by car, train, or bus, select an image from the road and try to remember a pleasant memory that's associated with that image. For example, a fast-food sign may enact a past memory of childhood, when you ate a meal at that restaurant. A tractor plowing a field may remind you of spending time at your uncle's farm. Try to challenge yourself with a new image every few minutes or so. It's a simple way to visualize old memories and a great way to pass time on a trip. If you're with a group of people, have them all do this interactively and describe your memories to each other.

Simon Says

Remember buying the electronic game Simon for your kids? Simon is a very simple computer game with four colored lights and four sounds. The game starts with one colored blinking light and one sound. You have to remember these and press that light and sound. As the game progresses, you must follow the increasingly complex visual and audio sequence.

This game strengthens your ability to memorize and visualize as you match your wit to Simon's challenge. A simple version of Simon can be found on the Internet at http://www.neave.com/games/simon/. You can purchase the Simon game at many toy stores for around $25.

Concentration/Solitaire

Concentration is an age-old game where you spread out a deck of cards, facedown, and then match pairs of the silly characters. With concentration, you have to memorize the location of the cards and then visualize this as you take your turn and "imagine" where the pairs might be.

Solitaire is yet another great neurobics exercise, as it requires memorization, visualization, strategy, and critical thinking. These games are perfect for car trips or while riding the train to the city for work. If you are at home, you can play these games free on many Web sites.

Memory Picture Drawings

Take a picture from your personal collection. Stare at the picture and visualize it in your mind. Now try to draw the picture on a blank sheet of paper. Can you do it? What features did you draw first? You may be surprised with your results!

An excellent follow-up to this visualization exercise is to put the picture underneath a tracing paper and draw the outline. This is good for your eye-hand coordination. Now try drawing the picture again after you traced it. Is your accuracy better? This is a challenge for your brain!

Living in the Big Brain Age

Nintendo DS games Brain Age and Brain Age 2 (http://www.brainage. com/) were developed by a neuroscientist and have many, many brain-stimulating games, from math questions to memory matching games. Led by Professor Kawashima himself (a computer image), these games will stimulate your mind and provide hours of entertainment. When I first tried Brain Age, I was stunned at the difficulty of the various challenges— timed math problems, timed memorization of lists of words, and more. Yet after doing this consecutively for several weeks, my score continued to improve; my brain age was actually getting younger, and I was enjoying it!

Big Brain Academy, also by Nintendo DS, is another superb brain-boosting neurobics activity that you can do alone or with others. Big Brain Academy, led by Dr. Lobe, a computer-generated professor, is available on the Nintendo DS Wii game set, and allows you to challenge others as you watch the game on your television set. Who knew that building a bigger brain and preventing Alzheimer's disease could be so much fun?

THINKING

Thinking neurobics should be done when you are able to take several minutes to be alone with your thoughts.

Read Shakespeare

Shakespeare employed functional shift—which means the great playwright sometimes used a noun as a verb or a noun as an adjective. This misuse of language actually insults the brain—and makes you think. You have to figure out what's going on, asking, "What did Shakespeare really mean?" Thus, reading Shakespeare forces you to determine the "sense" of the words—almost as if you had to solve an intricate puzzle.

Do Mathemagic

Math provides a great neurobics workout as you challenge yourself to various math problems. For instance, see how high you can count by twos, fours, fives, sevens, and nines. If that stops being a challenge, try counting in prime numbers, or in multiples of nineteen. Consider the square roots of large numbers. Or multiply three- or four-digit numbers such as 6,291 times 328. Now that should keep your mind occupied for a while!

Balance Your Checkbook

Toss your calculator for thirty minutes and balance your checkbook using your longtime ability to add and subtract (no fingers allowed). After you finish, check yourself with your calculator. If you are like many people at midlife, you've probably gotten a bit lazy when it comes to simple math.

Fuzzy Memories

Old memory activation provides a great neurobic stimulus. Lie down on your bed or hammock and feel relaxed. Now think back as far as you can to a distant memory of childhood—say the first day of school in elementary school. Now ask: What grade is this? Who was standing next to you? Was the teacher in the room? What was the teacher's name? How old were you in that memory? Next, categorize your memories by creating a mental time line—remembering times in first through twelfth grades. Were you able to re-create a memory of each year in your childhood and adolescence? With practice, you can do this—and you will be astounded at how it increases your ability to think on a deeper level.

REASONING

Reasoning neurobics increase your ability to interpret what's going on and verbalize this to yourself or someone else.

Foreign Films

If you have the means, watch a foreign movie without the subtitles. After watching a clip, see if you understand what's happening in the scenes, even if you do not understand the language. This is especially helpful in your communication reasoning.

Blind Taste Test

See if you can trick your mind. You will need a blindfold and five different types of drinks (cola, juices, dairy milk, soy milk). Have someone arrange several beverages in glasses while you are blindfolded. Then using your sense of *smell only,* see if you can determine the content of each glass.

Try this same blind taste test with different herbs and spices. Ask someone to place one teaspoon of each herb or spice on a tray—and mark this, so they know which is which. Now using your sense of smell, determine the names of these herbs and spices. A warning: Don't sneeze or inhale while doing this neurobics exercise.

Chess

Chess is a logic-based strategic board game and is also a fantastic reasoning neurobic. With today's technology, you do not even need a partner. Here's a great Web site I found where you can play against a computer: http://www.chessmaniac.com.

Mealtime Neurobics

Mealtimes offer an opportunity to bring all the senses to the table in a pleasurable and brain-healthy way. During a relaxing meal, you can engage in stimulating and challenging conversation with your spouse, children, friends, or coworkers, and these interactions have demonstrable positive effects on brain health. Also, by changing *how* you eat, without changing *what* you eat, you can boost your brain.

At mealtime, your visual, olfactory, tactile, taste, and even emotional pleasure systems are all in high gear, feeding associations into your cortex and tapping directly into the most primed memory circuits. The olfactory system can distinguish millions of odors by activating unique combinations of receptors in the nose. Here are some examples:

- Change the order in which you eat your food (dessert first!).
- Change where you eat your meal (dinner in bed).
- Use chopsticks instead of conventional silverware.
- Use the opposite hand for your silverware.
- Eat one bite of food at a time, savoring the taste, texture, and flavors.
- Try to capture a memory from the past (mashed potatoes remind you of Grandmother's kitchen at age six; the yellow mustard reminds you of playing Little League and eating hot dogs with mustard).

Engagement In Life

Being engaged with something meaningful in life is vital for experiencing optimal physical and mental health the rest of your life. Consider your options for volunteering if you enjoy helping people. If you are introverted, find classes you might enjoy. If you plan on retiring from work, determine where you will invest your hours of free time and energy. Whatever choice you make, making the commitment to fully engage yourself in a challenging and meaningful project or activity can help you continue to get the full mental benefit of building the brain's reserve. A good place to start is at one's church, synagogue, or mosque. Every city has service groups such as the Rotary Club, Kiwanis, Knights of Columbus, and many others where help is needed and new friends and projects abound. You don't have to be retired to join them.

Dr. Sears suggests that grandparents involve themselves in teaching and interacting with their grandchildren. Teaching them how to spell, add, and learn a foreign language are some examples. I'll always remember my grandfather teaching me Italian after dinner. I can still feel his beard against my cheek as he kissed me good night.

7

STEP 4: REST AND RECOVERY

Finding Your Circle of Quiet

My colleague, Dr. Marks, chair of the hospital ethics committee, never walked down the hospital corridor, he marched intensely. In fact, the sudden disappearance of every nurse on the floor was the preeminent alarm that always preceded Dr. Marks's presence. He was a perfectionist and demanded it of every person in the hospital.

One day a number of years back, I received a phone call from Dr. Marks's nurse requesting that I meet him at his office at 5:30 P.M. sharp for the annual USC ethics lecture.

Dutifully, I waited for the older doctor in the corridor outside his office when he came walking out (not storming out as usual). Dr. Marks seemed surprised to see me.

He asked me to drive and uncharacteristically he was silent during the thirty-minute trip to the university. While driving to the lecture, I remembered that I hadn't spoken with Dr. Marks for almost six months. When he did speak, I noticed that he had a certain indecisiveness and word-finding difficulty (called anomia).

At the lecture, as was his custom, Dr. Marks asked the poignant question on double effect in ethics, which he read from a scrawled note in his pocket. It was unusual for such a brilliant doctor to keep notes, I thought. Driving him home, I observed that Dr. Marks had at least ten folded notes stuffed in his shirt pocket.

More concerned than curious, I asked, "Did you think Professor Aaron's response to your question was adequate?"

Dr. Marks did not respond. However, when he did speak, it became apparent that there was a problem. He did not remember the question he'd asked the group, much less Dr. Aaron's response. This forgetfulness combined with his slower demeanor, pocketful of handwritten notes, and outward ambiguity in a normally precise and perfectionistic man was diagnostic.

Within weeks, Dr. Marks became my patient, and the origin of his problem was apparent. At age sixty-two, he was overwhelmed with enormous stress with a schedule that a roomful of interns could not do. After doing a physical exam, patient history, and laboratory tests, I diagnosed him with type 2 diabetes, hypertension, and the early stages of dementia. He also had extremely high levels of C-reative protein and homocysteine.

His wife had called me and commented on his snoring and moments when his breathing stopped during sleep, which she said had worsened in the past few years. She mentioned that he rarely slept for more than six hours at night and had not taken a real vacation since their honeymoon right before medical school.

In taking his family history, I found that Dr. Marks had no genetic predisposition for dementia. In fact, his parents lived well into their eighties with no sign of mental decline. Unequivocally, I determined that Dr. Marks caused his own mental decline, meaning his dementia was *lifestyle induced*. He had the two sentinel risk factors, stress and sleeplessness, and a clear case of sleep apnea. Dr. Marks was type "triple A" personality with the key traits of being keyed up or stressed out all the time; he was a sleepless workaholic and a solitary personality who took no time for reflection or social activities.

How about you? Will your fast-paced lifestyle and few hours of sleep each night result in Alzheimer's disease? I remember when I was confronted about going too hard and too fast about twenty years ago when our son, Mickie, age six at the time, bought two tiny hamsters at a local pet store. One hamster was busily eating her food while the other hamster ran constantly on the metal exercise wheel connected to the cage. Mickie could not stop laughing at the hyperactive hamster, going around and around, yet getting nowhere. I was laughing, too, until Mickie looked up at me and in all seriousness said, "Gee, Daddy, that hamster reminds me of you when you're all stressed up but have no place to go."

All stressed up but no place to go? Hasn't everyone felt that way? Sure there are times when we can all identify with running in circles like the hamster (or me!), trying to juggle myriad commitments in our lives with kids, careers, and other commitments. When you add caregiving to aging parents—especially to those with Alzheimer's disease—you have a guaranteed recipe for disaster . . . or as Dr. Marks experienced, a *recipe for early dementia.*

STRESSED OUT + SLEEPLESSNESS = DOUBLE DAMAGE

In this last step in the Anti-Alzheimer's Prescription, I want to delve into the double damage of stress combined with sleeplessness and how this increases the chances of Alzheimer's disease. I will discuss how chronic stress keeps the hormone cortisol at extraordinarily high levels in the body, which results in a catastrophe of hormonal imbalance, inflammation, and neurotransmitter dysfunction in the brain. I'll introduce new findings on sleep deprivation and how it increases pro-inflammatory markers in the body and may be to blame for the epidemic rise in obesity, type 2 diabetes, and Alzheimer's disease. I will also explain how inflammation and obesity are both linked to serious sleep disorders such as obstructive sleep apnea (OSA), a problem in which the person literally stops breathing (periods called apneas) many times throughout the night. OSA is increasingly common with aging, and is a risk factor for Alzheimer's disease. Finally, I want you to feel relaxed and get better sleep so you can enjoy social activities. Having a strong social network and developing religious or spiritual practices with a purpose-driven life are important ways to decrease the chance of Alzheimer's. I'll also show you how to optimize the neurortransmitters and receptor sites in your brain to keep your cortisol levels from wreaking havoc in your body and mind so you can maximize your ability to succeed under stress and reduce the risk of Alzheimer's.

Stress and Your Brain

Simply stated, stress describes the many demands—physical, mental, emotional, or chemical—you experience each day. It includes the stressful situation, or stressor, and the symptoms you experience under stress, the stress response. Stress can be negative (distress) or positive (eustress).

No one is immune from stress. We all experience it from day to day. Perhaps just hearing the word *stress* makes your head turn. Whether from an argument with your spouse, a confrontation with your boss, or fighting customers, clients, or traffic day after day during rush hour, *stress is real, and it's here to stay.*

Whereas physicians used to think that while stress made you feel uncomfortable, it wasn't really a big deal, we now know that daily stress can lead to tremendous emotional turmoil that shocks an immune system into a downward spiral, resulting in chronic or serious illness—and, yes, even increase the chance of Alzheimer's. In fact, some revealing findings indicate that the signs and symptoms of aging may be more related to our stress reactions than to our chronological age.

In research presented at the 114th Annual Convention of the American Psychological Association (APA), researchers explained evidence that biological and behavioral stress responses may be adaptive in the aftermath of stress, but can cause damage when they go on for a long time. In other words, whereas acute or short-term stress may enhance immune function and improve memory, chronic or long-term stress has the opposite effect; it can lead to cognitive impairment, as well as serious health conditions such as hypertension, depression, type 2 diabetes, and even cancer.

Stress And Coping Skills: Why We Respond Differently

What's most interesting about stress is the fact that none of us responds in the same way to stressful events. In other words, what may give you a feeling of emotional excitement might give your spouse or friend a sense of abject terror! That's because we all perceive and respond to stressors in different ways. However, it's the inappropriate responses to stress that influence your health and increase your chance of Alzheimer's disease and other illnesses.

The Austrian physiologist Hans Selye introduced the **General Adaptation Syndrome** when coping with stress. This means that the biological changes that occur in response to a stressor are beneficial because they enable you to adapt to the situation. This adaptation involves drawing upon resources within the body to provide the energy and oxygen that your body needs to either *fight or flee*.

Think of this process as being like your savings account at the bank. You set funds aside for use in an emergency. When your car unexpectedly breaks down, your life is barely interrupted because you can afford to have it quickly repaired. The emergency will not severely affect your lifestyle because you have the money to deal with it.

Your body's currency is stored away as triglycerides and glycoproteins. The currency itself is glucose and other sugars. This is what fuels your brain as well as the muscles and other organ systems within the body. Without it, your body would literally shut down.

Acute Stress Gives You Energy

So what happens when you find yourself in a threatening emergency? First, there's an immediate activation of the sympathetic nervous system and release of adrenaline. Then the brain sends a message to the adrenal glands. The primary mission of the adrenal glands is to produce a chemical or hormone that converts stored energy into usable energy. This chemical is the stress hormone cortisol, which I've talked about since chapter 1. Without cortisol, there's no way you could survive an emergency. Cortisol literally puts energy in your personal tank in the form of blood glucose.

When you have an acute emergency, one that lasts for a short time, no permanent physical damage is usually done. (In biological terms, a short time would be a few hours, perhaps even a couple of days.) Your heart rate and blood pressure increase and the alarm neurotransmitter adrenaline floods your body, making your heart beat faster and changing the blood flow to muscles and intestines. This adrenaline rush prepares you to fight the wild beasts (life's problems). Our built-in mechanism called the fight or flight response causes a profound set of involuntary physiological changes that allows us to handle acute stressful events. This response is controlled by the hypothalamus in the brain. When we face fear—or even recall a stressful or frightening event from the past—the resulting hormonal changes supercharge our body to a state of high arousal to prepare us for action.

Uncontrollable Chronic Stress Destroys the Brain

However, when stress happens for days, weeks, or months, it is called uncontrolled chronic stress. As an example, if you have a demanding job, financial problems, or a drawn-out divorce, chances are you live with chronic stress. Chronic stress occurs when we face stressors over a period of time. If you're a caregiver to an aging or ill parent, you may be experiencing uncontrollable chronic stress. Likewise, if you have a chronic illness like asthma, type 2 diabetes, or cancer, you may also have chronic stress. Uncontrollable chronic stress, in time, affects your body's genes by shortening the telomeres, the so-called tail of the gene. It has been shown in animal studies that stress decreases telomerase activity, which is important for telomere longevity. As I explained in chapter 2, telomerase is an enzyme that regulates how many times a cell can divide. Telomeric sequences shorten each time the DNA replicates. When some of the telomeres get too short, the cell quits dividing and ages. This is thought to cause (or at least contribute to) age-related problems like Alzheimer's.

When stress is chronic and the stress hormone cortisol is elevated over a period of time, it may shift fat distributions that precede many chronic diseases such as metabolic syndrome or insulin resistance syndrome (page 53). Abdominal fat or an increase in waist size also increases inflammation in the body, which is associated with an increased chance of Alzheimer's disease. Yes, that deposit around your waist is a toxic waste dump that fouls up the rest of your body.

When chronic stress lingers for weeks or months, problems arise. Because you are constantly withdrawing savings (naturally stored energy) without replacing them, you use up all of your body's resources until you reach a point where there are no more available. Instead of drawing upon naturally stored energy, your body will begin to break down muscle and other tissues to keep going. In addition, *chronic stress inhibits sleep,* which is the way your primary neurotransmitters regenerate. Without this regeneration, you'll feel fatigued, achy, overwhelmed, anxious, and depressed.

If you do not have a healthy way of responding to the chronic stress or counterbalancing the fight-or-flight response, the constant exposure to stress hormones will eventually cause your body to become overloaded and literally burnt out. When you're stressed out for a long period of time, it can result in a dramatic decline in both physical and mental health.

It's thought that stress causes an imbalance in our hormonal system and, in doing so, imbalances our entire neurotransmitter system. The increase of cortisol, the catabolic hormone, decreases lean muscle mass and increases fat and obesity. The increase of glucose sends insulin levels soaring. Over time, increased insulin levels increase levels of amyloid protein in the brain (the toxic protein that's the underlying cause of Alzheimer's).

We need some cortisol to keep our immune systems healthy. But when cortisol is elevated as with chronic stress, it can inhibit the lymphocytes or white blood cells or our immune system. Other chemicals produced by the brain's autopilot—the autonomic nervous system—can similarly damage the cells that comprise the immune system. Increased cortisol from stress increases the likelihood of colds, cancer, infections, and inflammation to the blood vessels and cells in the brain. Increased cortisol also increases our sympathetic nervous system and increases adrenaline and noradrenaline. After many years of being exposed to cortisol, a supercharged sympathetic nervous system can result in chronic fatigue and a decrease in executive function, attention, and concentration because of the decrease in blood supply through the vital hippocampus area of the brain. *Remember, the hippocampus is the part of the brain that's hit the hardest by Alzheimer's disease.*

During younger years, when cortisol is stimulated because of stress, there is a feedback mechanism that occurs in the hypothalamus of the brain that causes the cortisol to shut off. As we get older, this feedback loop does not work so well and hinders our ability to manage the stress response. Stress also interferes with weight loss and, most importantly, disrupts normal sleep through interference with the sleep cycle. *The stress/sleep double damage connection may be the greatest of all risk factor combinations for increasing the chance of Alzheimer's.*

DID YOU KNOW THAT . . .

- According to the American Stress Institute, stress is America's number one health problem, costing the economy more than $300 billion annually.
- Levels of the stress hormone cortisol are at their highest early in the morning, making you more likely to overreact.
- Watching the late-night news can elevate your cortisol and cause increased anxiety, making it tough to fall asleep.
- People who are anxious drink and eat more.
- Anxiety increases the chance of accidents, colds, heart attacks, and Alzheimer's disease.

The Signs and Symptoms of Chronic Stress

We have discussed the long-term effects of stress, hypertension, obesity, diabetes, and Alzheimer's disease. But what are the short-term signs of stress? It might be assumed that it is easy to know if you are stressed or not. The truth is, it isn't. An interesting study showed that over 50 percent of those tested did not recognize when they were under acute stress and especially did not recognize chronic stress. This was verified by me in a series of 100 patients I tested using biofeedback. Galvanic skin resistance was used to measure sympathetic activity, which increases under stress. The subjects, generally well educated, showed poor recognition of their stress in 40 percent of the cases.

The signs of stress differ from person to person. Generally, the neurotransmitters adrenaline and noradrenaline increase stress signs. The signs of relaxation, on the other hand, are triggered by other neurotransmitters such as serotonin and dopamine and the vagal system.

The acute physical signs of stress, from head to toe, are dilated pupils, dry mouth, and tense muscles, starting with the jaw muscles and then moving on to the trapeze, or neck and shoulder muscles. Breathing feels difficult and the increased rate of breathing can bring on hyperventilation followed by a panic attack characterized by shortness of breath and the hands and feet going numb. There is a sense of impending doom. The hands and feet become colder. Teeth clenching, especially at night, leads to worn tooth enamel. Other features of chronic stress are neck pain, tension headaches, and stomach ulcers.

Paying attention to your personal signs of stress is as important as paying attention to what you eat and how much sleep you get.

Why Sleep Is Important

To understand why sleep is so important, we need to go back to the concept of our hormonal symphony. Normally, we go through about six sleep cycles. There are also four stages of sleep, plus rapid eye movement (REM) sleep. The deepest stages of sleep (3 and 4) are necessary for the production of sufficient neurotransmitters, including dopamine and serotonin. REM sleep helps us organize and store our memories. Interruption of these stages is devastating to the brain and body.

Dopamine is the "get up" or pleasure neurotransmitter that motivates and promotes good habits. It's also linked to mood and plays a role in controlling appetite. When dopamine levels fall, the brain's ability to experience happiness and well-being suffers. People with less dopamine have a greater incidence of depression, addictions, and other neurological problems. Serotonin, the "I can do it" neurotransmitter, is relaxing and calmative, and helps us to sleep well and take things in stride. Serotonin also stimulates the hormone melatonin, which helps lull us to sleep at night.

Poor Sleep Increases Appetite and Weight

A recent study of healthy volunteers in the medical journal *Sleep* found that those who slept two to four hours a night were more than 200 percent more likely to be obese than those volunteers who got seven hours of sleep. In fact, one study found just a sixteen-minute loss of sleep per night also increased the risk of obesity.

These studies indicate that sleep loss lowers the level of leptin, a hormone that stimulates metabolism and decreases hunger. Sleep loss or shorter hours of sleep appear to boost the concentration of the hormone ghrelin, which increases hunger. In a controlled study, researchers have deprived healthy males of sleep and found that their levels of leptin went down, while ghrelin went up—both changes that increase appetite.

In line with these studies, there is increasing evidence that people who sleep less than six or seven hours a night have a higher risk for diabetes. Researchers at the University of Chicago found that losing just three to four hours of sleep over a period of several days was enough to trigger metabolic changes that are consistent with a pre-diabetic state (also metabolic syndrome). They determined that when sleep was restricted to four hours for six consecutive nights, the body's ability to keep blood glucose at an even level declined significantly. This may be because sleep deficit affects the immune function of the body. In one study, scientists found that a 45 percent reduction in total sleep time resulted in a nearly 30 percent reduction in cellular immunity. Getting quality sleep is now considered a basic defense mechanism to staying healthy and preventing disease. Getting seven to eight hours of sleep each night is your best defense against conditions such as obesity and diabetes that increase the chance of Alzheimer's.

Poor Sleep Reduces Levels of Human Growth Hormone

According to the National Institute on Aging, one out of eight individuals in their twenties has chronic insomnia, while one out of five people ages fifty to sixty-four, and one in four people over age sixty-five, experience this sleep disorder. Findings indicate that millions of women suffer with disordered sleep, especially insomnia. With age, the prevalence of insomnia increases as sleep time decreases, even though the time spent in bed might increase. An old adage to which women will attest is "When menopause begins, a good night sleep ends."

The problem with poor sleep is that it deprives the body of human growth hormone (HGH). Declining sleep quality begins between the ages of twenty-five and forty-five, according to a study reported in the *Journal of the American Medical Association.* In this study, researchers evaluated sleep study data from 149 healthy men, ages sixteen to eighty-three, and found that sleep deteriorates at two points in a person's life—between the ages of sixteen and twenty-five and again between the ages of thirty-five and fifty. Results showed that the time spent in deep sleep dropped from 20 percent (men under twenty-five) to less than 5 percent (men over thirty-five).

Recent studies from the University of Pittsburgh, which used functional MRI, indicate some patients with chronic sleep disturbances have activity in central sleep and alerting areas of the brain that should be shut down as they attempted to sleep, but were not. Instead, in these individuals with long-term sleep disturbances, the areas were activated by attempting to fall asleep. Identifying these patients is important to correct this particular sleep disorder.

Obstructive Sleep Apnea and Alzheimer's Disease

While snoring is caused by the vibration of the soft parts of the throat while breathing in and out during sleep, obstructive sleep apnea (OSA) involves periods of breath holding while snoring. The periods of stopping breathing (called apneas) are caused by obstruction of the upper airway. Apneas may be interrupted by a brief arousal that does not awaken you completely—you often do not even realize that your sleep was disturbed. Yet if your sleep were measured in a sleep-disorders laboratory, technicians would record changes in the brain waves that are characteristic of the arousals.

The most brain damaging effect of OSA is that it interrupts sleep stages 3 and 4 and REM sleep. In fact, in severe OSA, a person may never reach stages 3 and 4 at all during the night. I first realized that I had OSA when my transcriber told me I had fallen asleep during the dictation. When she played back the tape, you could hear me talking for ten minutes, and then nothing but my snoring and breathing interruptions. That was a wake-up call for me! I diagnosed at least two patients a month for ten years before I recognized I had it.

Obstructive sleep apnea results in low oxygen levels, which result when the blockages prevent air from getting to the lungs. The low oxygen levels also affect brain and heart function. OSA is more common than asthma in adults, and up to two-thirds of those with OSA are overweight. Those with more than twenty apneas (complete obstructions) per hour of sleep may have a greater risk of dying from cardiac rhythm and rate disturbances and complications of high blood pressure, such as stroke and heart attacks, than people with fewer apneas.

Treatment of OSA involves weight loss and sometimes using CPAP, a device that creates positive airway pressure. Studies show that treating sleep apnea properly with positive airway pressure leads to lower blood levels of C-reactive protein and levels of two markers of platelet activation.

If you wonder if you have OSA, ask your spouse if you snore, or tape-record yourself while sleeping. If you are fifty, overweight, male, and a snorer, chances are you have OSA. Women are not immune to this disorder either. Please don't delay. Talk to your doctor about OSA, as it is a parasite that robs the vitality of your body and brain.

A Self-Fulfilling Prophecy

When we cheat ourselves of ample sleep—and the normal cycles of sleep—a self-fulfilling prophecy occurs. Less dopamine is secreted, which causes our willpower to wane. Less dopamine leads to

- a bigger appetite and binge eating or drinking;
- a decrease in leptin (a special hormone that makes you feel full);
- an increase in ghrelin (a hormone that stimulates hunger and appetite); and
- an increased tendency for depression.

Less sleep also results in decreased amounts of serotonin. Having less serotonin alters your body's biorhythms, increasing your sensitivity to pain and resulting in early morning awakening with feelings of dread, alarm, and panic. Less serotonin also increases the appetite, especially for simple carbohydrates (chips, cookies, cakes, candy, white bread, and pasta), which I believe is the direct root of the obesity epidemic in the United States today. (On a lighter note, my practice dietitian reminded me that *stressed* is *desserts* spelled backward!)

With less serotonin, we require more drugs to fall asleep, and these chemicals interfere with our own biorhythms and hormonal balance. With the decrease in dopamine, our executive functioning decreases, as does our ability to follow through with plans and commitments. In addition, because of decreased dopamine, our self-control markedly diminishes. I've done myriad informal studies on hundreds of my patients through the years, and I've found that those individuals who suffer with addiction problems—whether alcohol, drugs, sex, or food—also have very low dopamine receptors in their frontal lobes and other important places in the brain.

Prescription Steroids, Sleep, and Stress

There cannot be a discussion about sleep and stress without talking about the diurnal cortisol cycle and the effect of prescription steroids.

The stress hormone cortisol has a biphasic cycle that is high during daylight hours and low at night. At approximately 6:00 to 8:00 A.M., levels of cortisol begin to rise in the body. At about 9:00 to 12:00 P.M., levels of cortisol rapidly diminish and stay at low levels until the next morning. This decrease in cortisol level allows us to sleep. Low cortisol levels are associated with an increased production of the neurotransmitters dopamine and serotonin, chemicals that are necessary to deal with the next day's stress. Anything that disturbs this diurnal cycle and turns cortisol "off" will disturb sleep and decrease our ability to handle stress.

Now let's add prescription steroids such as Prednisone, Medrol, and Dexamethasone, as well as the over-the-counter hydrocortisone creams, eye ointments, and eye drops. It is estimated that more than 2.8 billion doses of Prednisone and other steroids are given in pill form, just in the United States. The problem is these steroid preparations increase levels of cortisol, sometimes up to a hundredfold, putting our bodies and brains in an alarm state. These drugs interfere with sleep and prevent the production of dopamine and serotonin. In addition, doses of steroids increase the blood glucose level, which causes an increase in insulin or type 2 diabetes, and an increase in blood pressure—all risk factors for Alzheimer's disease.

Before using steroid medication, you must cautiously weigh the risks and the benefits. Always consult your doctor. Do not use over-the-counter drugs or creams that contain these cortisol-like steroid compounds. I believe these drugs are a direct cause of the increased incidence of early onset Alzheimer's in the United States.

ALL STRESSED OUT AND SLEEP DEPRIVED

After months of poor sleep and chronic stress, you feel exhausted with a low mood and no energy. You probably binge on simple carbohydrates (chips, candy, desserts) to satisfy the dramatic decline in dopamine and serotonin. The increased carbs cause insulin, your symphony conductor, to go wild, with its levels soaring upward. When insulin soars, your limbic (emotional) brain takes control, and your neocortex (thinking) brain takes a backseat. You may also try to jump-start your body by doing the following:

1. Bingeing on carbohydrates;
2. Getting high or numb with sleeping medication and alcohol;
3. Drinking tons of coffee for a jolt; and
4. Adding salt to food.

All of these are abusive! If you continue this body/mind abuse regularly, it becomes a destructive habit. Over a period of months, the destructive habit becomes a full-blown addiction. Common addictions include nicotine, alcohol, and sleeping pills, both over-the-counter and prescribed. Less significant addictions include caffeine and salt. Thus, the key reason I call stress and sleeplessness the double damage that increases your risk of Alzheimer's is because the loss of dopamine and serotonin cause key hormonal changes that result in a direct inability to cope with life's stressors. With the double damage, your hormonal symphony is definitely imbalanced and out of key.

The limbic brain eventually leads you into a disharmonious life filled with bad health habits, eating disorders, addictions, obesity, metabolic syndrome, type 2 diabetes, and eventually Alzheimer's disease. Continue to keep in mind that 70 percent of the time, Alzheimer's is *not* genetically programmed—you control your destiny, and following the four steps in this program will let you stay Alzheimer's-free. In the other 30 percent—those cases that are genetically linked—we can still delay the onset of Alzheimer's by more than ten to fifteen years by watching our diet and lifestyle habits, including controlling stress and getting quality sleep. It is never too late—or too early—to control your stress response and increase your quality of sleep.

Granted, the older we get, the harder it is to sleep. Older adults have less deep-sleep time, more arousals and disruptions during sleep, and less efficient sleep. When missing a few hours of sleep over a week continues for several weeks or months, the damage to your body is cumulative, for it results in markedly increased inflammatory cytokines, which increase systemic inflammation. So how do we stop the cycle?

HORMONAL BALANCE AND PSYCHOLOGICAL RESPONSES

To reduce the chance of Alzheimer's disease, you must seek hormonal balance. Unbelievably, the stress hormone cortisol actually helps you to achieve that balance. Because cortisol is intimately regulated by the limbic (emotional) brain, emotional stress can, through cortisol, have a profound impact upon anabolic and catabolic hormones.

During stress, there are many hormones produced by the body, including some, such as growth hormones, that actually augment the immune system. Whether the immune system will be affected when you're stressed is determined in large part not by the concentration of any one hormone, but instead by the relative amounts of several different ones.

Furthermore, these hormones are profoundly influenced by your psychological response to stress. Let's look briefly at these five responses:

1. Perception

Of the four psychological responses, your *perception of the stressor* is absolutely paramount. Unless you think something is going to be dangerous, it is not going to trigger the so-called alarm phase of the General Adaptation Syndrome, the stage when you realize that there is a potential threat to your well-being.

What is the difference between deleterious or uncontrolled stress and normal, everyday concerns? A lot! Sure, we are all busy. We all have concerns and responsibilities. Nevertheless, stress occurs when you feel something is way *out of your control*, that it's far more than you can handle. Also, with stress, you might fear the consequences or what "might" happen.

Uncontrollable stress, then, is a combination of being unable to cope because of a situation we cannot control, along with the resulting fear. Is it true that there are some who imagine that they are under stress and others who actually are under stress? The answer is yes. Unfortunately, it is your perception of the stressor (and imagined consequences) that is often the determinant.

Consider this scenario. As you get into the elevator on the 22nd floor of your office building, you calmly push the button to go down to the parking garage. The elevator begins to descend normally, and then suddenly it jerks, makes a loud squeaking noise, and stops. All the lights go out. In just seconds, you've gone from feeling calm and in control to feeling trapped—with no control. Your limbic emotional brain jolts into overdrive and the immediate alarm response is the outflow of adrenaline triggered by the sympathetic nervous system. Your heart races. Your muscles feel tight. Your respiration becomes rapid and deep. In simple terms, you are ready for action! At the same time, another stimulus has gone to your pituitary gland to excrete a longer-term response to this immediate threat. This stimulus causes the adrenal glands to release the stress hormone cortisol.

As you frantically look for the control buttons, the elevator jerks, and the doors suddenly open. At this moment, another physiological response occurs. This time the parasympathetic nervous system calms and balances the sympathetic nervous system (your alarm system). The parasympathetic nervous system now mobilizes its internal defenses and causes a soothing excretion of hormones that relax your body and, in particular, protect the hippocampus, one of the central portions of the limbic system and source for memory transmission.

Now—unbeknownst to you—there is a raging war going on between your brain's alarm system and its calming system to reach a state of balance, or *homeostasis*. At this moment, the body's stimulating and tranquilizing chemical forces are clashing. The sympathetic nervous system and parasympathetic nervous system are at a tug-of-war with each other. Your hormonal symphony is out of harmony.

Problem is, for some individuals, these calming chemicals never occur, and there is a constant stimulation of the alarm system. I treat many of these individuals after they lose it completely and eventually have nervous breakdowns.

When the alarm system is triggered by an external event such as the sudden stopping of the elevator, the production of anabolic hormones, such as growth hormone, and reproductive hormones, such as testosterone, estrogen, and progesterone, is decreased. Even the immune system goes on hold and blood flow to the skin is reduced. These physical reactions are why stress leads to decreased sexual function and an increased risk of illness like asthma, allergy, hypertension, diabetes, and cardiovascular disease.

In addition to these risks, when the alarm system is triggered, your blood pressure and heart rate are increased, and blood vessels are constricted because of an increase in epinephrine. The oxygen-enriched blood going to the brain increases initially because of the increased respiration and heart rate caused by the fight-or-flight reaction. Vast amounts of energy are available because the adrenals have caused a rapid release of glucose and fatty acids into your bloodstream. In addition, your senses become more acute, with your pupils dilating to give you greater peripheral vision. Your sense of smell becomes highly acute, and, initially, even your ability to think and your memory are vastly improved. You also become less sensitive to pain.

The best way to turn off the body's alarm system is by physical activity, as described in step 2 (chapter 5). Physical activity burns up the excess glucose and adrenaline and, in fact, helps metabolize it back to normal levels. However, if there is no physical activity, as in this case, being stuck in an elevator, the high glucose levels lead to an increase in insulin. If glucose levels are increased for a long period of time, it can decrease brain glucose, causing agitation. The high insulin levels cause a decrease in the removal of amyloid proteins from the brain—and, as I've said, increased amyloid proteins in the brain is a key sign of Alzheimer's disease.

2. Control

When you feel you have no control, emotional excitement, which is positive, turns into anxiety, which is negative. There is a very fine line between the two conditions. It is similar to the fence that separates you from a caged tiger at the zoo. The same animal ten feet from where you are standing is simply an object of curiosity when there is a protective barrier. The moment that barrier is removed, the tiger becomes a threat. Being in control is similar to that protective barrier. It is when the sense of control is lost, and you feel helpless, that anxiety and subsequent health problems can occur.

3. Coping Style

Coping style is another psychological variable that can influence your health. For example, I know several emergency room doctors who spend hours each day dealing with horrific tragedies and making life/death decisions for people they've never even met before. Yet, in the midst of that incredibly stressful environment, they function admirably and are in excellent mental and physical health. The stress does not get them down. I also have dear friends who've made their fortune, retired early, and have nothing more to do each day than to sit in a chair and wonder "if and when" their children will come to visit them. Ironically, these people have the most difficulty coping with life's stressors.

Animal studies have clearly shown that some people can actually develop coping mechanisms that can function even when they feel they have no control. An example of one such study used rats in a maze. When the rats were shocked, their cortisol levels increased and continued to remain high as long as the rats perceived they had no control. However, once the rats were given a lever to push that would prevent the shock, their cortisol levels returned to normal or diminished.

This study showed that even when the lever was no longer preventing the shocks, the rats "thought" that they were being protected, and so the cortisol levels did not elevate as much as they had initially. The researchers concluded that the same outcome might be true with human beings. We'll explore coping mechanisms more later in this chapter.

4. Personality Type

Finally, personality types can have a profound impact upon susceptibility to illness. I mentioned that Dr. Marks was a type "triple A" personality. You have probably heard of the type A personality, individuals who are quite literally slaves to the clock. They are always in a hurry and seldom take time out to enjoy life. They are time-oriented and frequently speak with a very rapid rate of speech, finishing the sentences of those they are speaking to, as they grow impatient. This is in marked contrast to the type B personality, who is quite simply a non–type A. The latter individual gets the work done, but he also takes time out to enjoy the journey of life.

While this is a useful way to categorize people, it might be more accurate to think of "personalities" as coping styles that enable people to deal with problems in their immediate environment. These are just a few of the many factors that can affect both hormonal balance and the immune system, especially the way they respond to stressors.

5. Life Balance

Another way I have patients manage stress is to evaluate their life balance. In other words, is their life stable or lopsided? An unbalanced life is a major cause of stress for most baby boomers. I give my patients the life balance test. I ask them to consider eight categories and rate them from 1 to 10: 1 being they are very deficient, 10 being they feel very good about that category, 5 being they believe it is adequate. A life balance rating of 5 to 10 is considered a good rating; less than 5 is a deficiency.

How do you feel about your . . .

1. Diet
2. Exercise
3. Spiritual Life
4. Sex Life
5. Social Life
6. Job or Occupation
7. Marriage or Significant Other
8. Sleep

After they have rated these 1–10, I ask how much time they devote to each. This is often a reality check when some find that 90 percent of their time is spent at work and there is no time for anything else. I advise that anything under 5 needs work. A perfect 10 on everything is a sure sign of an unrealistic person with a problem, like Dr. Marks. A patient who scores less than 5 on four or more of the life balance signs is in danger of chronic stress.

Preventing Alzheimer's with Rest and Recovery: *Finding Your Circle of Quiet*

Both rest and recovery are crucial to balancing the body's hormonal symphony. By rest, I mean getting quality sleep every night, making an effort to relax your body and to get in bed early enough to get seven to eight hours of sleep. By recovery, I mean the ability to place the body in a state where the adrenaline sympathetic nervous system is reduced and the vagal, or relaxation, nervous system increased.

There are many tools you can use for recovery, including the relaxation response, deep abdominal breathing, listening to music, meditation, prayer, yoga, and saying the Rosary using prayer beads, among others. In each of these relaxation remedies, you stimulate the frontal inferior temporal gyrus, the "optimistic" center of the brain. During recovery, your blood pressure and pulse decrease, your skin warms, and your muscles relax. Using relaxation techniques frequently throughout the day for recovery helps to increase dopamine and serotonin levels, which, in turn, enable you to sleep deeper and longer. We've seen the results of relaxation techniques repeatedly in the Buddhist monks who can stay alert for twenty-four hours at a time and continue to function normally.

In one important meditation study performed at the University of Wisconsin, researchers were able to show changes in the brain using a functional MRI (FMRI) when different groups of people meditated. For instance, when the veteran monks meditated, the FMRI showed a large area of the frontal temporal optimistic lobe light up. Laboratory results indicated that dopamine and serotonin levels were also high at this time. When the novice monks meditated, they did not have the same level of the light-up; however, the FMRI indicated there was more activity than normal in this brain area. When college students meditated, the FMRI showed small areas light up on the scan. Researchers determined that by constant reinforcement (continuing to meditate frequently every day), the pathways for this optimistic center of the brain are increased to all areas of the frontal lobe and they can be turned on more easily. With meditation, the pessimistic center of the brain is suppressed, so negative thoughts are diminished.

I believe this is also why those individuals who are religious or who pray or meditate daily have the lowest incidence of Alzheimer's disease and also increased cardiovascular health—for they have less panic and adrenaline and consequently lower levels of insulin in the body and brain. With higher levels of dopamine and serotonin in the body, we don't procrastinate and we resolve inner conflicts, which controls our alarm reaction. Because we feel good and in control, we can rest calmly and get deeper sleep. We do not escape to addictive food or carb cravings, or binge on alcohol or drugs that artificially calm us or stimulate dopamine pathways and lead us to deplete the supplies of serotonin and dopamine.

Let the following suggestions help you *find your circle of quiet* as you increase rest and recovery and regain control of your emotional state.

1. Identify and Eliminate Stressors

To reduce stressors, you must identify the major stressors in your life, such as problems with money and relationships, grief, and deadlines. If you can't resolve these stressors alone, get professional help for problems that seem too difficult to handle.

Also, never hesitate to say no before you feel overextended with too many commitments. Respond instead by saying, "Let me think about it." Or say, "Let me get back to you." Especially if you are balancing a career, along with raising kids and other volunteer commitments, you should not feel guilty about prioritizing what is humanly possible. Take time weekly to evaluate your commitments and do only those that are most important, saying no to the remaining tasks. Saying no, when appropriate, can bring your stress to a manageable level and give you some control over your life.

Realize that it's okay to be "good enough." When the pressure cooker of life begins to explode, remember that you are *one person*. We can do the best possible or be good enough, but we also have to realize our humanness and allow for this.

2. Talk It Out

Talk to a friend, family member, mental health counselor, your pastor, priest, or rabbi, if your stress level is too high. Getting your feelings out without being judged is crucial to good mental health. As a rule of thumb, psychological counseling can help you to develop coping skills, so that life's stressors do not overwhelm you.

3. Take Time Out

Before you reach your breaking point from life's unending stressors, take a time-out for solitude. Being alone does not mean feeling lonely, for we can feel lonely in the midst of a crowd or even sitting with our family and friends. Being alone can help you find your circle of quiet in life— that inner place that brings you meaning in life. Take time to nurture yourself away from the cares and responsibilities of the world and find time for inner strength and mind, body, and spiritual healing.

4. Rekindle Your Spiritual Side or Religious Beliefs

There are hosts of studies showing that people who focus on spirituality or attend religious services are better able to handle life stressors and less likely to get Alzheimer's disease. Religion can give a sense of purpose and feelings of hope, even when life seems hopeless. Religious involvement also provides good social support essential for well-being, and encourages a more compassionate, forgiving world. Findings show that when we feel contented, supported, and forgiven, levels of cortisol are lowered in the body.

Volunteering to help others through your religious organizations or community benevolent groups is yet another way to rekindle spirituality and feel more connected to God or a higher power as you stop dwelling on your own problems and focus on giving to others.

Prayer and spirituality proclaim forgiveness, which is associated with an essential equation that is not directed by any neurotransmitter. It is a *decision* that helps to make us civilized, that is, the ability to help others and have compassion rather than participate in the barbarism of the strong taking advantage of the weak. A study at Hope College in Michigan showed a 30 percent decrease in mortality in those who learn to forgive. Physiologically, forgiving turns off the right inferior temporal lobe pessimistic center and turns on the optimistic center, triggering dopamine and serotonin, which in turn stimulate our anabolic hormones, putting our symphony back into perfect harmony.

5. Strengthen Your Social Support

Connections to a partner, family, friends, or a support group have been shown to improve mood and ability to cope and can even strengthen your immune system. Most people who are able to cope with stress have strong social support networks with family, friends, and even pets.

A strong social network also has an impact on Alzheimer's. Studies have found that people with active social networks are less likely to develop the disease. In a 2006 study, researchers examined the brains of people with Alzheimer's who had recently died with the characteristic plaques and tangles of distorted protein. The researchers also had data on cognitive symptoms and how sociable the individuals had been during their lives. Researchers concluded that even among those with extensive plaques and tangles, Alzheimer's symptoms were less severe if the people had many friends.

6. Learn to Meditate

Brain scans have revealed that meditation, a highly active mental state, produces a mental condition somewhat similar to non-REM sleep (which many specialists believe is the more mentally restorative sleep phase). This is believed to occur because the many neurons of the cortex fire in harmony during meditation. Unlike sleep, consciousness is fully maintained in meditation, so there is no grogginess upon awakening. In addition, meditation has been found to surpass all forms of relaxation therapies at lowering blood pressure, as reported in the journal *Hypertension*.

With meditation, you may focus your mind on one thought, phrase, or prayer for a certain period of time. You pay complete attention to what is happening in the present moment without being distracted by what has already happened or what might happen. When you do this, it leads to the relaxation response, a physiological state that helps to decrease heart rate, blood pressure, respiratory rate, and muscle tension. Meditation also decreases hormones such as cortisol and adrenaline, which are released during the fight-or-flight response.

Meditation can guide you beyond the negative thoughts and agitations of the busy mind and allow you to become "unstuck" from your fear and other disturbing emotions. Once you've learned how to meditate effectively, you can switch into this relaxation state at will—before stressors cause you to feel overwhelmed.

How to Meditate

The following are two types of meditation, one passive or centering meditation, the other active or process meditation. The first clears the mind, the other controls the optimistic process.

Allow fifteen to twenty minutes a day to see benefits.

1. Sit in a comfortable chair in a quiet room. Make sure there are no distractions. Close your eyes as you begin to meditate.
2. In centering meditation, you focus your attention on the repetition of a word, sound, phrase, or prayer, doing this silently or whispering. An alternative is to focus on the sensation of each breath as it moves in and out of your body.
3. Every time you notice that your attention has wandered (which will occur naturally), gently redirect it back, without judging yourself. If you continue to practice this, you will learn how to do it correctly.
4. In process meditation, you listen to an inner voice that frees you to view the good things that happened that day and let go of experiences that were negative. Some people find they are able to plan and prioritize their future by listening to an inner voice.

7. Practice Relaxation

Relaxation helps to increase the body's morphine-like pain relievers—endorphins and enkephalins—which are associated with a happy, positive feeling. Relaxation therapy may also improve your quality of sleep. One small study of several different relaxation procedures found a 42 percent improvement in self-reported sleep complaints after one year of relaxation therapy.

The relaxation response is a physical state of deep rest that changes the physical and emotional responses to stress. This physiological state is inborn in all of us and can occur at times when you are not aware of it.

How to Relax

Set aside a period of about twenty minutes that you can devote to relaxation practice. Remove outside distractions that can disrupt your concentration.

1. Lie flat on a bed or floor, or recline comfortably so that your whole body is supported, relieving as much tension in your muscles as you can.
2. During the twenty-minute period, remain as still as possible; focus your thoughts on the immediate moment, and eliminate any outside worries that may compete for your attention.
3. As you go through these steps, in your own way try to imagine that every muscle in your body is now becoming loose, relaxed and free of any excess tension. Picture all of the muscles in your body beginning to unwind; imagine them beginning to go loose and limp.
4. Concentrate on making your breathing even. As you exhale, picture your muscles becoming even more relaxed, as if you somehow breathe the tension away. At the end of twenty minutes, take a few moments to study and focus on the feelings and sensations you have been able to achieve. Notice whether areas that felt tight and tense at first now feel more loose and relaxed, and whether any areas of tension or tightness remain.

I am often asked, "How do you know when you are successfully meditating and relaxing?" During my years at the Institute of Living hospital, we studied biofeedback that examines the body's physiological signs during the state of relaxation, including alpha brain waves, galvanic skin resistance, and hand temperature. We found that a certain percentage of people did not know when they were truly relaxed. However, by making these people aware of bodily signs that indicated a truly relaxed bodily state, they could determine when they were, in fact, relaxed.

How will you know you're relaxed?

1. Increased moisture in the mouth
2. Warmth in the hands and feet
3. A feeling of heaviness in the limbs, neck, and lower back
4. A warm tingling of the skin and a feeling of well-being, even euphoria, which means dopamine is working in the body.

KIDS & ALZHEIMER'S PREVENTION

Teaching a child to pray and have faith is a strength that might help them through their most difficult times in life. My father "Ole Sox" used to say, "You cannot protect your children from life; you can only prepare them for it." Whether through prayer or living an exemplary life, adults must teach children how to prepare for the years ahead.

---

8. Laugh More

Norman Cousins was the first to promote laugher as an antidote to disease. While serving as editor of the *Saturday Review*, Cousins was diagnosed with a serious connective tissue disease. Although his doctors said that it was incurable, as Cousins shares in his book, *Anatomy of an Illness as Perceived by the Patient*, he was determined to get well. While bedridden, he watched funny movies and used laughter to create positive chemical changes in his body, which resulted in improved health.

Learning to laugh more and worry less can give you great benefit with rest and recovery. During stressful times, rent some funny videos and watch these instead of the nightly news. You'll sleep better after a good laugh, which gives your entire mind and body a healing boost. An interesting Canadian study found that even looking forward to laughter— or anticipating fun—can help boost the immune system and reduce stress. In the study, researchers tested sixteen men who all agreed that a certain video was humorous. Half of these men were told three days in advance that they'd watch this video. Those who knew this in advance began to experience biological changes immediately. Then when the men actually watched the video, levels of cortisol declined by 39 percent in their bodies. Levels of adrenaline also fell by a startling 70 percent and endorphin levels (the feel-good hormone) rose 27 percent. Human growth hormone levels climbed 87 percent! A practical place to start is to examine the movies you watch. Do they stir up fright, provoke conflict, or make you smile and laugh and give you hope and peace?

9. Exercise Daily

If you are a chronic worrier, I urge you to get up and start exercising more often—and regularly. A chronic worrier is someone who worries about every situation in life. You can block the effect worry has on your health by doing something positive. Not only will exercise ease your emotional anxiety and reduce dangerous levels of cortisol, it will boost serotonin and dopamine and help to boost blood flow throughout your body, resulting in optimal health and healing for your immune system. I find that going for a walk after eating a meal allows me time to sort through my thoughts and relax.

10. Let Music Soothe Your Soul

Studies show that listening to soothing music lowers blood pressure and boosts immune cell count while reducing levels of stress hormones. Avoid melodies that make you tense or that cause uneasiness. Spend ten to twenty minutes a day listening to music, and try this in combination with another relaxation technique such as deep abdominal breathing or while walking outdoors or on a treadmill.

How to Do Deep Abdominal Breathing
1. Lie on your back in a quiet room with soft music playing in the background.
2. Place your hands on your abdomen, and take in a slow, deliberate deep breath through your nostrils. If your hands are rising

and your abdomen is expanding, then you are breathing correctly. If your hands do not rise, yet you see your chest rising, you are breathing incorrectly.

3. Inhale to a count of five, pause for three seconds, and then exhale to a count of five. Start with ten repetitions of this exercise, and then increase to twenty-five, twice daily.

11. Get Healing Sleep

If you have difficulty sleeping because of too much stress, consider the following sleep suggestions:

- Meditate or pray right before bedtime to calm your mind and body.
- Sleep only as much as you need to feel refreshed, but no more.
- Avoid daytime napping if you have trouble sleeping at night. However, you might try a brief twenty- to thirty-minute power nap (a siesta), which many people find refreshing, especially between 2:00 and 4:00 P.M. when your sleep hormones are peaking.
- Wake up at the same time every day, weekday or weekend. This strengthens your circadian cycle—our daily rhythmicity—and will help to establish regular sleep patterns.
- Use earplugs if you are bothered by noises while sleeping. Some people find that "white noise"—a machine that produces a humming sound or turning the radio to a station that has gone off-air—helps.
- Reduce light, especially alarm clock and television lights if they flicker or pulse.
- Avoid caffeine after 12:00 noon each day.
- Avoid alcohol, as it produces a light, fragmented sleep.
- If these therapies do not work, talk to your doctor. Sometimes non-habit-forming sleep medications may be helpful in easing you to sleep. But don't become dependent on them.

> If you find yourself taking more than seven sleeping pills per month repeatedly, see a sleep specialist. Most sleep medications interfere with natural biorhythms.

12. Ask Your Doctor If You Might Need Medication

On a side note, I believe that antidepressants such as the selective serotonin reuptake inhibitors (SSRIs) may decrease the chance of Alzheimer's, although no studies have shown conclusive evidence. The SSRIs elevate serotonin levels in the brain while also suppressing dopaminergic pathways, thus helping to resolve problems with stress and sleep. If you have difficulty relaxing and trouble sleeping, an SSRI may help you sleep deeper and have clarity of mind. This, in turn, allows you to make better lifestyle choices and deal with stress in a much healthier manner. Personally, I believe that antidepressants are excellent medications to help reverse the two sentential risk factors, if patients are truly anxious or depressed and need this medication. Not all antidepressants facilitate sleep, so ask your doctor which is best.

DIAGNOSIS, TREATMENT, AND THE FUTURE OF ALZHEIMER'S DISEASE

PART III

DIAGNOSIS, TREATMENT,
AND THE FUTURE OF
ALZHEIMER'S DISEASE

8

IS IT ALZHEIMER'S?
MAKING THE DIAGNOSIS

Last summer, George, a seventy-two-year-old former CEO, came to see me, accompanied by his wife and two daughters. In his own limited way, he let me know that he was there only because his family "tricked" him into coming. He inferred that his wife and daughters were convinced that he could no longer take care of himself and now they would take his money. Twice during his rambling dissertation, George stumbled to find the right word for "tricked." First saying "kicked," then "licked," he then exclaimed, "You know . . . fooled." After many more attempts, the correct word, *tricked,* finally was spoken.

In an effort to calm the man down, I touched George on the shoulder and asked him if he knew my name. "Sure, you're a white coat, a medicine man. Your name is . . ." he repeated five times. To put an end to this frustration, I said, "My name is Dr. Fortanasce." He retorted, "No, that's not it!"

The family reported that they had taken George to their primary care doctor five times in the past year. At the first visit, George refused to let anyone go with him to the doctor's appointment. His wife called the doctor to explain her concerns, but the doctor merely said, "Oh, it's nothing. We all get older. George is just feeling his age."

The second time (after several of his personal checks bounced), George was accompanied by his wife. This time the doctor did an EKG, a chest X-ray, and a complete blood count. All tests were negative. The doctor advised George to get a financial adviser.

By the third visit to his primary care doctor, George had become withdrawn and had little interest in playing golf or in other activities. His wife said, "I know it's his age, but he is so irritable now."

So the good doctor started George on an antidepressant to lift his mood and said he'd see him again in two months. Each subsequent visit seemed shorter than the one before and consisted of taking George's vital signs and listening to his heart and lungs.

Weeks later, George's wife called the doctor and told him her husband had worsened. He was throwing temper tantrums and accusing her of not putting things where they belong. "He's confused and claims that I'm cheating on him with this man in the corner of our bedroom," she whispered on the phone so George would not hear.

This time, the primary care doctor prescribed Thorazine, an antipsychotic medication, and told her that George might need to see a psychiatrist if he continued this behavior.

Soon after that, George began to feel faint when getting up from a chair. Then, one day while his wife was combing his hair, she noticed a painful, red lump on his crown. George didn't remember falling. Again, she called the primary care doctor, but he was out of town, and the covering physician said, "I don't know George's history. You might go to the ER, if you're that concerned."

By this time, they called me *without* a referral from their primary care physician—simply because one of their daughters had gotten my name from her own doctor, and she insisted that her parents see a neurologist.

A thorough neurological exam revealed severe short-term memory loss with good long-term memory. As an example, George could name every player on the 1928 New York Yankee baseball team, yet he didn't know the current date or year. I also found some paranoia such as his believing "this is not really my wife but an impostor."

This paranoia, called *Capgras Syndrome*, is a common occurrence with Alzheimer's. The person no longer believes the husband or wife is their spouse. Hearing George say this aloud caused his wife to break down crying right there in the examination room. (This is a frequent and heartbreaking occurrence with Alzheimer's patients and their spouses.)

The neurological exam also showed severe vibratory loss of both lower extremities, which is a sign of B_{12} deficiency and common in el-

derly adults. Correcting this problem is usually as easy as boosting vitamin B$_{12}$ supplementation. George also showed a right Babinski reflex, a
sign of a left hemisphere injury, which was indicated through his impaired balance.

Perhaps the most stunning finding was George's blood pressure,
which was 140/80 lying down but only 90/50 standing up—a reading
that's dangerously low for an elderly person. That day, I performed a lab
and imaging. George's lab test confirmed what I thought—a critically
low vitamin B$_{12}$ level of 180. His MRI scan showed a small, but definite,
subdural hemorrhage.

Once I got George in treatment, which included brain surgery for
the subdural, he improved greatly. His daughter called to report on his
improved mental status, but added, "He's still not the same dad I've always known."

Could George have avoided some of these problems if his family had
seen a neurologist early on when he first displayed symptoms? Certainly.
Allow me to show you why finding the right physician and obtaining
proper diagnostic tests are so important early on when symptoms of Alzheimer's first begin. Having the right treatment early could save you or
your loved ones from a misdiagnosis, incorrect treatment, and unnecessary family stress.

Of course, in this part of the chapter I will speak to you, the reader,
with instructions on how to plan for a time in your life when you might
have problems with memory or Alzheimer's and what you should expect
from your physician. Perhaps you don't need this instruction at the present time. That said, you could use these criteria as a plan for selecting a
physician for a loved one—a parent or even your spouse—who may have
signs of Alzheimer's.

Selecting the Best Doctor to Treat Alzheimer's Disease

Most physicians can do the initial workup of a patient suspected to have
a memory problem. Most physicians can come to a preliminary diagnosis. That said, however, Alzheimer's is the great masquerader and certainly not an easy diagnosis to make in the initial stages (like mild

cognitive inpairment, discussed in chapter 1). Only certain physicians can make this diagnosis accurately.

What You Must Do

If you suspect that you have memory problems, it's important to ask someone close to you—your spouse, a child, or a colleague—to accompany you to the doctor's visit and be there with you during the exam. When you make the initial appointment, it's imperative that you explain the problem and ask for *an extended visit.* Helping your doctor understand that this entails a memory problem will enable him or her to select the proper tests and focus the exam.

An early diagnosis is important and often mind-saving for the following reasons:

1. It allows for timely intervention for treatable dementia, as discussed on page 223. Some problems associated with treatable dementia include vascular problems (vascular dementia), vitamin B_{12} deficiency, incorrect medications or combinations of medications, drug or alcohol problems, or other untreated medical problems.

2. It allows for intervention that might reverse the development of the dementia. Sometimes simply changing a medication, changing the strength of a medication, or adding vitamin B_{12} supplementation can correct problems associated with early dementia.

3. It allows for the patient and family to prepare for the future. If the problem is, in fact, Alzheimer's disease, there are medications available to help with the early symptoms. (I review these medications in chapter 9.) These treatments might improve the patient's quality of life in the early stages and also allow time for the family to adjust to what may lie ahead with Alzheimer's disease.

4. It is essential in the recognition of mild cognitive impairment. The risk of MCI developing into Alzheimer's increases at a rate of up to 18 percent per year. The Anti-Alzheimer's Prescription is of particular importance to this group.

Is This the Right Physician for You?

When you're considering a physician to make an accurate diagnosis of Alzheimer's, it is important to select someone who is trained and experienced in treating brain diseases. Not all physicians are equal in their knowledge of a particular disease.

Be aware that problems often occur when there is more than one doctor administering treatment. For instance, you might have a primary care physician, an allergist for problems with asthma, a cardiologist for hypertension, and even a rheumatologist for problems with osteoarthritis. Unless effective communication takes place between all the physicians, you are putting your health in danger. As an example, your primary care physician may prescribe an antihistamine to help control a runny nose. Yet when you develop an upper respiratory infection, your allergist may write a prescription for the antibiotic erythromycin. Sometimes a combination of medications such as this is potentially dangerous and can cause irregular heart rhythms and other serious problems. While both doctors are working at making you well, it is important to have *one doctor* who knows all about you—your various health conditions, the signs and symptoms, the tests you've had, the diagnosis, the treatment plan, and all medications and natural dietary supplements you take. It must be emphasized that a knowledgeable patient is one who reports all the details about his or her life—including changes in medications—to each doctor.

SPECIALISTS WHO MAY TREAT ALZHEIMER'S DISEASE

Primary care physician: This doctor is a general practitioner, a family practice doctor or pediatrician, or internist who has completed three years of training after medical school graduation.

Geriatrician: Geriatricians are doctors who specialize in care for people sixty-five and older. These doctors are typically board-certified in internal medicine and have additional training in areas pertaining to elder care.

Psychiatrist: Psychiatrists are medical doctors who specialize in the evaluation, diagnosis, and treatment of mental disorders. Psychiatrists can prescribe medications, and, in addition, psychiatrists may treat people through counseling.

Neurologist: Neurologists are doctors who specialize in diseases of the brain and nervous system. These doctors are best suited to diagnose and treat Alzheimer's. They have extensive knowledge of all nervous system disorders and often have good training in recognizing psychological problems. They are considered the best to diagnose and treat Alzheimer's disease.

Neuropsychologist: A neuropsychologist works with a neurologist in performing psychological testing that might identify cognitive problems.

Age, Sex, Credentials, and Coverage

In choosing a health care professional, some people ask friends for recommendations, check the physician's credentials, or call the local hospital for referrals. In this age of managed care, you will need to check the list of doctors who will accept your insurance provider. Nevertheless, none of these methods is foolproof in finding a qualified professional with whom you can feel comfortable to share innermost feelings and concerns about memory problems.

Perhaps one of the most important steps to take when selecting a physician to diagnose and treat a memory problem is to know yourself—including your personal likes and dislikes. As you go through the process of choosing a physician, consider the following twenty questions. Some of these questions will pertain to your initial selection of a physician. Others are to consider after you've seen this physician several times—just to help you make sure this is the right doctor for you.

1. Would you feel more comfortable with a man or woman?
2. Should the physician be older than you, the same age, or younger?
3. Do you have a preference as to educational background?
4. Is the doctor board-certified? This means that the doctor passed a standard exam given by the governing board in her specialty.
5. Where did the doctor go to medical school? Your local medical society can provide this information.
6. Is the doctor involved in any academic pursuits, such as teaching, writing, or research? This doctor may be more up-to-date in the latest developments in the field.
7. Where does the doctor have hospital privileges and where are these hospitals located? Some doctors may not admit patients to certain hospitals, and this is an important consideration for older adults with other health problems.
8. Does the doctor accept your particular type of health insurance, or is the doctor a member of the medical panel associated with your HMO?
9. Is the doctor's staff friendly and reassuring? Do they smile and make you feel valued? Chances are the staff reflects the personality of the physician.
10. What are the doctor's office hours? Are these hours convenient for you or someone who is transporting you?
11. During the initial visit, does the doctor go over a thorough review of your history, including medications, past surgery, lifestyle habits, and family history of Alzheimer's disease?
12. Does the doctor greet you looking at you—like you are a person of value?
13. How much time does the doctor spend on follow-up visits? Does the doctor allow you time to tell your story?
14. Does the doctor examine you with a mental status exam if memory is the problem?
15. Does the doctor order tests readily, or does he or she tend to minimize your concerns?

16. Is the doctor ready to give you a prescription without explaining more about the side effects?
17. Does the doctor return your phone calls?
18. Does the doctor make you feel that your health comes above all else? Alternatively, do you fear your health care plan dictates the quality of care you receive?
19. If you need hospitalization, will this doctor still treat you? Or will you be delegated to a "specialist hospital doctor" who knows nothing about you as a person? *You must ask the doctor this question!*
20. Does the doctor use specialists to assist with your situation, if you request one? Sometimes health plans discourage physicians from referring to other specialists. Or the physician may have bad rapport with other specialists. Both are warning signs to find a new physician or get a new health plan.

On a side note, because of managed care, finding the right person to diagnose and treat Alzheimer's disease properly and cost-effectively is not always easy. For those with a health maintenance organization (HMO), a "gatekeeper" or primary care physician must make the referral to the neurologist. Carefully read the policy manual to understand the specific rules, and then select a physician whom you can trust to know your personal and family medical history and take responsibility for your health care.

Write Down Concerns and Seek Answers

Before your appointment with the neurologist, write down a list of concerns you may have about memory loss and Alzheimer's disease and specific symptoms you might have. It is also helpful to get an in-depth family history before meeting with your doctor. So often, a family history of Alzheimer's is crucial in making an accurate diagnosis and prescribing effective treatment.

On a side note, I mentioned "brain freeze," which invariably occurs when you fear the worst and your adrenaline and cortisol levels peak and overwhelm the receptors. That's why it's important to always be prepared before you sit in front of the diagnostician and report on symptoms. Before your visit, consider and record the following:

- Your mental and physical health concerns
- Symptoms you've noticed, such as forgetfulness or memory glitches
- Symptoms others have noticed about you (family members, colleagues)
- Unusual behaviors you've exhibited
- Your health history, including conditions such as hypertension, type 2 diabetes, cardiovascular disease, and other problems that may increase the risk of Alzheimer's
- Your family history of Alzheimer's disease
- Medications you are taking now and in the past, including prescribed and over-the-counter; unusual side effects of medications you are taking or have taken
- Natural dietary supplements you are taking
- Your lifestyle habits (exercise, diet, smoking, alcohol consumption, drugs such as marijuana, cocaine, amphetamines, tranquilizers)
- Your sleep habits
- Causes of stress in your life (marriage, work, social)
- Questions you have about Alzheimer's disease

Review Your Medications and Supplements

When it is time to see the doctor, bring your medications and any nutritional supplements you may be taking on your first visit. Your doctor will let you know which ones are safe to keep taking, depending on an analysis of the drugs or supplements and your new diagnosis and medications.

So if you wonder how you can be sure your doctor will take you seriously about memory loss or fear of Alzheimer's, remember that the squeaky wheel often gets the grease. If you know what to ask, you will get what you need! Be organized before your visit and be open and honest with your doctor at the visit, and you will find answers. Something else: Always ask questions. If your doctor seems annoyed at your questions, that's a red flag and you should look for another doctor!

Does Anyone Really Know Elizabeth?

Elizabeth, seventy-nine, came to see me with her youngest daughter, Kimberly, who had just flown in from New York City. Kimberly had not seen her stepmother in several years and could not give an adequate history to me, and Elizabeth was too busy twisting and turning and mumbling to know why she was sitting in my office.

It was obvious that Kimberly was bothered by having to take her stepmother to the doctor. She continuously received calls from clients on her cell phone, even during the brief appointment. Finally, when she finished a lengthy phone conversation, she spoke. "Okay. I think we're here because she's not getting good sleep and my dad is worried. In addition, she's a bit forgetful. Then, again, Elizabeth's never had a much of a memory to speak of."

How do you begin to make a diagnosis without pertinent facts about the patient? So I began by asking some key questions:

Was Elizabeth on any medications?
What dietary supplements did she take?
Who was Elizabeth's primary care doctor?
Did she have other health problems, such as cardiovascular disease, type 2 diabetes, or hypertension?
Was there anything in particular about which Elizabeth had been confused recently?
Why didn't her husband accompany her, so I could have more patient history?
Had Elizabeth wandered off or forgotten where she had parked her car?

My list of questions about this petite elderly woman continued.

After thinking about all these questions, Kimberly finally spoke. "Gee! When I think about all your questions, it makes me wonder if I have a memory problem. I often forget where I park my car in the city, and last week I missed a hair appointment."

Sadly, no one really knew Elizabeth—nor did they want to know

what was really happening in her mind. Because her family was not in tune with Elizabeth's mental problems, her diagnosis and effective treatment were delayed. Make sure you and your family talk openly about Alzheimer's disease. My hope is that it never happens to you or your loved ones. But should you become aware of early signs of Alzheimer's in yourself or your spouse, it's vital to talk about the changes and see your doctor. When you come to the doctor, bring the person who knows you best, so they can add insight into new changes of mental functioning. By taking advantage of early diagnosis and proper treatment, many people with early Alzheimer's can continue to live meaningful lives.

Making the Diagnosis: What Doctors Must Do

All doctors must be extremely thorough in making an accurate diagnosis of Alzheimer's disease. Here are the some pertinent steps that I follow with every new patient:

- **Take the patient's medical and family history.** In cases of Alzheimer's, a person who knows the patient well and has recent knowledge of their mental state must accompany the person. Sudden deterioration versus a slow progressive change is important for the physician to know in order to decide what tests to order and what treatment to give. Tests that should be done are the MMSE (Mini-Mental Status Exam), the clock drawing tests. Go to anti-alzheimers.com for examples of testing.
- **Record vital signs.** The doctor will take the body temperature, pulse or heart rate, blood pressure, and respiratory rate to assess the most basic body functions.
- **Check vibratory function in the feet.** With this test, the doctor uses a tuning fork to check the patient's vibratory reflex on the knee and big toe. Vibratory sensation is part of a system of pressure sensation in our feet that tells us if our weight is on our toes or heels. It's the information needed for our gyroscope located in the brain stem. Without this information, the gyroscope cannot keep us upright. For example, when the lights are out and we cannot use our vision for

balance, it's this sensation that helps us keep our balance. It's the major reason why older people fall at night or when they're on uneven surfaces.

- **Test for the presence of Babinski's reflex by scratching the bottom of the feet.** Babinski's reflex is one of the infantile reflexes that are normal in children under two years old, but the reflex disappears as the child ages and the nervous system becomes more developed. In those over age two, the presence of a Babinski's reflex indicates damage to the nerve paths connecting the spinal cord and the brain (the cortico-spinal tract).

- **Test for the Snout Reflex,** the Glabella Tap, and the Palmar Mental Reflex. As discussed in chapter 1, these Alzheimer's tests elicit reflexes normally seen in newborn infants that help them root, suck for feeding, protect their eyes, or bring food to their mouths. The "grasp" reflex, another test, usually reappears in severe Alzheimer's. The grasp reflex is often misinterpreted as sign of violence or stubbornness, as the Alzheimer's patient will grab on to your hand or arm and not let go until they are distracted with another object to cling to.

Mandatory Lab Tests

If you have a possible memory problem, the doctor must order specific laboratory tests in order to make an accurate diagnosis and then treat the problem. Trust your doctor to decide which set of tests is best in your case to ensure no other medical problems are present. This can help you avoid extra testing that may add little to your diagnosis and only increase the number of tests and expense. If you fear one specific diagnosis, such as brain cancer, be sure you tell your doctor. If you still do not feel comfortable with the diagnosis, talk to your doctor and then have more testing. Or it's always your right to get a second opinion until you have peace of mind that the problem has been diagnosed correctly. Then— and only then—can proper treatment and healing begin.

Complete Blood Count (CBC). Your doctor may get a sample of blood for a complete blood count and chemical profile. These results will help assess your general health and eliminate any other disease as a possible cause of the memory loss problem. This test measures the amount of red and white blood cells and shows how your vital organs such as the kidney and liver are functioning.

C-reactive Protein. C-reactive protein is a marker of inflammation in the body. Levels of C-reactive protein are elevated during infections, and people with heart disease and those who are obese have elevated levels. The development of future heart disease is often predicted by elevated levels of C-reactive protein. As I've explained, we now associate pro-inflammatory markers in the body, including C-reactive protein, with Alzheimer's disease.

Homocysteine. Homocysteine is a product derived from the metabolism of methionine, an essential amino acid predominant in animal protein. Studies show that, at high levels, homocysteine damages artery walls, which can cause cholesterol to build up and block the vessels. High levels of serum homocysteine is also correlated with cognitive dysfunction. Taking supplements of B vitamins (folate, B_6, and B_{12}) may lower elevated homocysteine levels. Eating foods high in folic acid, and vitamins B_6 and B_{12}, may also help lower homocysteine.

Vitamin B_{12}. B_{12} (cadalmadim) is a catalyst for normal red blood cell production and neural function. This lab is best done with a fasting blood test. Deficiencies are usually due to an inherited factor and sometimes common in blond-haired blue-eyed people. A vitamin B_{12} deficiency can cause neuropathy in a severe treatable dementia, which is usually seen after age forty.

Folic acid. Folic acid is the co-enzyme for purine synthesis required for nucleo protein synthesis and blood production. A deficiency in folic acid can cause a problem similar to B_{12} deficiency.

TSH or Thyroid panel. A thyroid panel will show the level of thyroid hormones. A level below the normal range may cause depression. Thyroid is also decreased with stress, pollutants and pesticides, and people who have a history of "yo-yo" dieting. Drugs such as lithium also decrease thyroid production. Other deficiencies such as iodine, tyrosine, selenium, and manganese can also negatively affect thyroid production.

Symptoms are fatigue, low body temperature with an increased sensitivity to cold, weight gain, constipation, depression, and treatable dementia. Patients with underactive thyroids may need primarily to take thyroid hormone medication. The thyroid is often compared to the idle of a car's engine. If the car's idle is too low, the car sputters. If your thyroid hormone is too low, your brain function sputters.

Complete metabolic panel. This battery of blood chemical tests includes liver enzymes, electrolytes, and kidney function (blood urea nitrogen or BUN, serum creatinine) and is done to understand disease states and the function of organs.

Additional Laboratory Testing

Additional lab tests are necessary only if there is clinical suspicion for specific types of dementia. These tests should include calcium, phosphorus, zinc, magnesium, copper and ceruloplasmin cortisol, human immunodeficiency virus (HIV), antiphospholipid antibodies, and antineuronal antibodies. While these tests mean nothing to the patient, they are important to a physician who knows how to diagnose dementia.

Lumbar Puncture

A lumbar puncture or cerebral spinal fluid (CSF) collection is a test to look at the fluid that surrounds the brain and spinal cord. Cerebral spinal fluid acts like a cushion, protecting the brain and spine from injury. While this procedure is usually not needed in those patients under sixty-five years old, it is advisable. This procedure may be important if the younger patient has a history suggestive of hydrocephalus, infections, vasculitis, or cancer. The fluid is examined for cells, protein, and glucose, as well as for infections and inflammation such as syphilis, and HIV. Tests for markers of Alzheimer's disease such as tau and amyloid-beta peptide are becoming standard.

Electroencephalography (EEG)

The EEG uses an apparatus for recording electrical activity from the brain. It uses special electrodes or probes placed on the scalp attached by wires to an amplifier that can convert the electrical signals to wavelike written forms on papers or as images on a computer screen. The EEG can help the doctor determine memory loss associated with an alteration in consciousness (such as seizures).

Polysomnography (Sleep Study)

If you snore or suffer with daytime fatigue, you may need a polysomnography (sleep study), which is done in a special laboratory. This sleep study includes a recording of electroencephalography (EEG), electrooculogram (EOG), and electromyogram (EMG) in order to assess the actual quality of your sleep. These tests provide important data that define the time it takes you to fall asleep, the duration of your sleep, and the time you spend in the different stages of sleep. Brief arousals, full awakenings, and movements are recorded to determine the severity of the fragmentation of sleep, which might account for daytime sleepiness and other symptoms.

All of the tracings from the EEG, EOG, EMG, and the respiratory monitors are carefully reviewed, literally second by second, by a trained technologist and a specialized physician. These tracings will help your doctor determine the quality and quantity of sleep, the continuity of airflow at the nose and mouth, and the movements of the abdomen and chest wall. The number of minutes of sleep is counted and the percentage of time spent in each stage is calculated. These specialists count every arousal, awakening, and movement, along with every apnea and hypopnea. The amount of time spent at various oxygen levels is also determined.

THREE TESTS YOU MIGHT NEED WITH A SLEEP STUDY

Electroencephalography (EEG) uses an apparatus for recording electrical activity from the brain.

Electrooculogram (EOG) records the electrical voltage that exists between the front and back of the eye.

Electromyogram (EMG) involves an instrument that converts electrical activity associated with functioning muscle into a written or visual record.

Necessary Imaging Tests

Though imaging techniques are equivocal in early diagnosis of Alzheimer's, these tests are most important in ruling out treatable causes such as hydrocephalus, tumors, and vascular disease. Imaging tests may suggest the diagnosis of Alzheimer's when atrophy is noted in the frontal and temporal lobes.

Structural Scans: CT and MRI

To determine the cause of memory problems, computerized tomography (CT) or CAT scans or magnetic resonance imaging (MRI) of the brain is mandatory. By giving a more detailed image of the brain, these scans may be able to confirm memory loss and be predictive of who might develop Alzheimer's. The MRI scan currently offers the most sensitive, noninvasive way of imaging the brain. The structural MRI can assess the shrinkage of the brain, especially in the hippocampus. Certain medical centers can now quantitate hippocampal atrophy or loss. This has been correlated with mental status changes to predict those who have a high probability of developing Alzheimer's disease. As mentioned, we call this mild cognitive impairment.

Functional Scans: PET and F-MRI

Both the Positron Emission Tomography (PET) and Functional Magnetic Resonance Imaging (FMRI) are sometimes used to assist with a diagnosis. The PET scan shows reduced brain cell activity in certain regions of the brain, which may be consistent with the diagnosis of Alzheimer's disease. The FMRI is an MRI scan that tests the metabolism of the brain and shows which neurons are "on" and which ones are "off." The FMRI is critically important for a greater understanding of brain physiology, especially that dealing with accentuating the brain's reserve.

When Should a Specialist Be Called?

If the memory loss is associated with walking difficulties or any localized weakness, loss of vision, sudden speech loss, loss of consciousness or sudden hallucination, illusions, or severe confusion, your doctor may want to consult with another specialist on your case. Also, if there's no clear explanation from lab tests or the EEG, then I recommend an immediate specialist evaluation. If the memory loss is progressive over an extended period of more than three months or remains steady for three months, a specialist's consultation is warranted. Most importantly, if your physician does not do the minimum history, physical examination, and laboratory and imaging tests, you must seek another opinion.

Treatable Dementia Versus Diseases That Contribute to Alzheimer's

Treatable diseases are those caused by factors other than degenerative amyloid plaque and tangles or those due to genetic factors. For example, remember George? I diagnosed him with a vitamin B_{12} deficiency. Vitamin B_{12} is essential for neuronal vitality and life. If not for a lack of vitamin B_{12} in his diet, he probably would not have had the memory loss. But vitamin B_{12} deficiency is more common among elderly adults, and it's a problem that must be considered (and then treated) when making a diagnosis.

Giving George vitamin B_{12} supplementation intramuscular through the skin, which bypasses the gastrointestinal tract, restored normal function to those cells that had not already died. By not diagnosing a vitamin B_{12} deficiency in time, irreparable damage can occur. Along with the vitamin B_{12} deficiency, George also had a recently occurring massive brain hemorrhage, which worsened because of the intense swelling and pressure. Removing the subdural hemorrhage, giving vitamin B_{12} supplementation, and alleviating George's severe hypotension (low blood pressure) by changing his medications ended his signs of memory loss or dementia. Strokes can also cause vascular dementia. But vascular dementia is not the same as the dementia caused by Alzheimer's disease.

Is It Alzheimer's or Not?

Some of the signs and symptoms of an illness such as major depressive disorder or bipolar disorder may mimic Alzheimer's, particularly the symptoms of depression, confusion, insomnia, paranoia, and hallucinations. Sometimes physicians get confused while trying to differentiate the diagnosis. However, the patient with Alzheimer's disease also has other more noticeable symptoms, including difficulty performing familiar tasks, problems with abstract thinking such as balancing a checkbook, impaired memory and forgetfulness, an inability to follow simple commands, problems with language and communication, and poor hygiene, among others.

Wanda became concerned about her mother, Rose, age seventy-two, whom I had treated for major depressive disorder for almost two decades. Wanda called me one evening and said she'd been at her mother's home about an hour away and noticed some dramatic changes in her habits, finding "a carton of eggs and package of lunch meat" in her mom's warm kitchen cabinet. She checked other rooms and found a once-frozen pizza stored neatly on the shelf in the laundry room, next to the detergent. Wanda said that her mother had been forgetful lately and had even forgotten her name several times when she called to check up on her. "Several times when she went to the hairdresser or to meet girlfriends for lunch, she forgot how to get home and a police officer had driven her home," Wanda said, "and when I'd suggest that she talk to her doctor, she'd become angry at me."

I urged Wanda to be more assertive and bring her mother to see me for testing. Within days, Rose, who had suffered with major depression for twenty years, was diagnosed with the early stages of Alzheimer's disease.

Currently, there is *no* definitive diagnostic test for Alzheimer's disease, but a probable diagnosis is obtainable. But Alzheimer's disease is not the same as major depressive disorder, which can happen at any age. Feelings of depression can often coexist with a physical illness.

For instance, Frank, age seventy, came to see me with symptoms of low mood and forgetfulness. In doing the physical examination, I routinely ordered a Prostate-Specific Antigen (PSA) test, which measures the level of PSA in the blood. PSA is a biological marker or tumor marker and can help detect disease. When Frank's PSA test came in moderately elevated, I referred him to a urologist at UCLA, who later diagnosed Frank with prostate cancer. Within a week, he underwent surgery, which was successful in treating the cancer. When he came back to my office two months later, his depression and forgetfulness had resolved.

Many times patients with underlying diseases such as prostate or pancreatic cancer present with depression as the main symptom. Patients with chronic obstructive pulmonary disease (COPD), including emphysema and chronic bronchitis, may have depression and difficulty sleeping. That is why it is imperative to see a medical doctor immediately for a complete evaluation if you or a family member has signs and symptoms of depression, moodiness, forgetfulness, or other similar problem. The mood change may be a red flag alert to another serious health problem.

Attributable Causes of Alzheimer's

As I've continued to explain throughout this book, I firmly believe that certain risk factors increase the chances of getting Alzheimer's disease—whether obesity, hypertension, abnormal lipids, type 2 diabetes, chronic stress, sleep disorders, or environmental factors. These factors either cause Alzheimer's, or they can cause Alzheimer's to progress much faster than it normally should in those who are genetically predisposed.

For instance, some people are genetically programmed to get vascular disease. They can eat a good diet, exercise, and do all the right things, but by age fifty, like everyone else in the family before them, they get occlusion of an artery. Others who are *not* genetically predisposed can also suffer an MI (myocardial infarction or heart attack) by age fifty solely by eating a diet high in trans fats, saturated fats, and sodium, along with living a sedentary lifestyle. The same is true for Alzheimer's disease. It is a man-made, lifestyle disease—and *we can only blame ourselves*. However, sometimes in a radical attempt to solve a problem, we create another more complex problem. Take my patient Sophia as a clear-cut example.

Sophia's Weight Loss Surgery and Dementia

Sophia, a once-successful trial attorney, came to see me. At fifty-nine, her legal career was failing, and she had been asked by three senior partners to take a leave of absence. They told Sophia that she was making far too many mistakes and forgetting important assignments and appointments. But what really concerned her partners, causing them to take drastic action, was when Sophia fell asleep in court during an important trial. When the judge asked Sophia, "Do you want to ask any questions to the witness?" her response was nonexistent, for she was sound asleep.

Like many women, Sophia came alone to my office and openly said she was having some memory problems. I noted that Sophia had a history of obesity and hypertension, but at the time I saw her, she was only mildly overweight.

The correct diagnosis was not made until I asked the nurse to assist her in undressing for the exam. When I returned to the room and started listening to her heart and observing her body, I saw a longitudinal, relatively recent, surgical scar across her abdomen. I determined immediately that Sophia had recent gastric bypass surgery, which she confirmed. We're just learning more about gastric bypass surgery and how it's becoming widely known to cause vitamins B_6, B_{12}, and E, iron, and other deficiencies that can lead to dementia, imbalance, and neuropathy.

Sophia had another problem that was never diagnosed until my office visit with her. She had obstructive sleep apnea (OSA). Sleep apnea, as I discuss in step 4 (chapter 7), is increasingly common in obese men over age fifty, and in women after age sixty. The problem with sleep apnea is that it can lead to brain deterioration in the hippocampus, the learning and memory center of the brain, among other serious health problems, even stroke.

Sophia's recent memory loss and her difficulty walking compounded other risk factors leading to signs of early dementia and general nervous system breakdown.

Sophia told me that she saw her primary care doctor and a cardiologist every three months. Somehow, both of these physicians had missed these symptoms, and thus Sophia's personal life and career suffered greatly. If the problem is recognized early, the treatment is simple. In Sophia's case, it was as simple as supplements of vitamins B_6 and B_{12} and iron tablets, along with a breathing device called the CPAP to wear at night. The nasal CPAP has a mask, which is strapped on your nose, connected to a swivel and flexible hose to a special pump, which quietly provides air under pressure to your nose. The CPAP maintains a positive pressure inside your airway while you breathe. It acts as a support to prevent further narrowing or collapse of your airway, and it actually increases the size of the airway behind the palate and at the back of your tongue. Problem is, so many older adults like Sophia are not diagnosed in a timely manner and are just another new statistic as the more than 650,000 Americans under age sixty-five with Alzheimer's or other dementia grows.

The Worried Well

No chapter on the diagnosis of Alzheimer's can be complete without discussing the "worried well." The first hint that a physician is dealing with a worried well patient is when they make a statement similar to the following: "Doctor, I have serious memory loss and let me give you the exact details."

First, if you have memory loss, you don't remember the "exact details"! Many of the worried well patients come alone or their spouse sits patiently as they elaborate at length—and in detail—about their illness.

As an example, Alice, seventy-four, came to see me about her serious memory problems. Active in several volunteer organizations, Alice maintained a lovely home and even worked part-time at her daughter's clothing boutique in Los Angeles. Alice said, "Dr. Fortanasce, my memory problem started exactly two months and five days ago. I remember the moment as if it were yesterday. A police officer stopped me for speeding when I was coming home from work. When he approached my car, I blanked out. I couldn't remember where I put my driver's license or insurance card. The police officer was a dark-haired, muscular gentleman, and he assured me that everything was fine."

Alice continued, "Since that time, my memory has failed me several times. For instance, sometimes I can't remember my children's names. I call Janice 'Joan' and Joan 'Janice.' I know I'm losing my mind."

Alice needed an immediate evaluation, and I was happy to give her a clean bill of health! Another woman, Meredith, age forty-five, gave a similar history of periodic memory loss to me. But Meredith also had a focal neurological symptom: peripheral vision loss. I ordered an MRI, which showed a pituitary tumor. After treatment for the pituitary tumor, Meredith's memory problem ended. There is no room to discuss the complete differential diagnosis of Alzheimer's disease.

Though I do not expect you to read about all the myriad Alzheimer's mimics, I do hope I've impressed upon you the complexity of the human brain and the absolute need for a competent consultation.

WOMEN AND ALZHEIMER'S DISEASE

If you're a woman reading this book, you may wonder if your feelings of memory loss or forgetfulness could be related to Alzheimer's disease. Perhaps your husband has noticed that you were forgetful and you attributed it to not getting enough sleep or to menopause. I have found that many women hesitate in asking their doctors about symptoms of Alzheimer's, thinking the doctor will not take them seriously.

This is a growing concern of women—and physicians: that doctors take them and their symptoms seriously. Some experts believe that women are simply less aggressive than men are in demanding treatment; others believe that many health care providers regard women as hysterics and feel they overuse the health care system more frequently than men.

While this overutilization of health care services may have positive results with prevention of disease, some physicians believe the overutilization is obsessive and linked with female "psychological issues." For instance, in one medical center, there were reports of 75 percent of women with endometriosis being dismissed after multiple treatment failures. While all of the women had verified chronic pelvic pain, their physicians virtually wrote them off as overutilizers of the health care system and dismissed their pain as being "neurotic." In another survey, women with fibromyalgia syndrome (FMS), a chronic pain–related ailment, averaged seeing four to five physicians before receiving a proper diagnosis. Because there are no laboratory tests for FMS, the doctors told the women the pain and symptoms were "in their heads."

No matter what anyone tells you, *Alzheimer's is very real.* The signs and symptoms of memory loss are not imagined. If you notice that you're more forgetful than normal or start having episodes of memory loss or feel highly anxious or depressed, call your doctor immediately. Don't slough it off as being stressed or tired or think it's related to aging. Talk with your spouse or another family member about your concerns, and make an appointment to see your physician with this family member.

Early treatment may help you to feel normal again. If the symptoms are determined to be Alzheimer's disease, then you and your family members need to take this seriously and discuss the future.

The Schulman Study Gives Insight

Perhaps the Schulman study explains why women's complaints are often underdiagnosed and treated inadequately. This comprehensive report published in 1999 in the *New England Journal of Medicine* studied the relationship of race and gender to physicians' recommendations for managing chest pain. The results of the study were quite revealing and may apply to any medical problem, including Alzheimer's disease.

The study investigators concluded that physicians assessed female patients as being less intelligent, less self-controlled, and more likely to overreport symptoms than male patients. While this bias was *not* intentional, it often results in misdiagnosis and undertreatment for women. In addition, for women who may suffer with anxiety, depression, or signs of memory loss, undertreatment of these types of problems sometimes means no life at all.

I urge you to love yourself enough to call your physician if you have signs and symptoms of Alzheimer's disease—or any illness. If your physician does not take your symptoms seriously, please find a doctor who does.

Is Your Brain Really Healthy?

If you remember nothing else from this book, please remember this: *Early diagnosis of Alzheimer's disease gives a better chance for treatment.* If you have a reversible memory problem, taking care of it quickly when treatment is most effective is the best step you can take. If you do one thing for yourself, you owe it to yourself to be certain that your brain remains as healthy as possible. The Academy of Neurology estimates that 75 percent of patients with clear evidence of dementia are not diagnosed in a manner timely enough to treat them.

That said, by making smart diet and lifestyle choices, you can begin to slow the degradation of the brain and, in effect, slow down your body's biological clock. I believe this is an exciting promise. To be honest, this is the *only hope* we have right now for preventing Alzheimer's disease.

9

THE LATEST MEDICAL THERAPIES
FOR ALZHEIMER'S DISEASE

The high-pitched screams echoed like a siren off the cold white tiles covering the emergency room walls. As I entered the small examining room, Mrs. Santos, a petite Hispanic woman in her late sixties, strained and struggled and tried feverishly to bite at the restraints on the gurney. Her pupils dilated with a frenzied stare, and she ignored the pleas of her two daughters, who begged her to calm down, to relax.

The oldest daughter spoke first and told me there had been a dramatic change in their mother's personality and behavior over a six-month period. "Her personality changed just after surgery for breast cancer," the daughter said. "She refused radiation and other treatment. And the doctor said that probably no other treatment was needed.

"Then Mother began complaining of numbness in her arms and legs," her daughter continued, "and shortly after this, she started to act peculiar, very paranoid and suspicious, and she even hallucinated once."

The daughter told me that Mrs. Santos's primary care doctor had done an MRI of her brain a week ago, and it was negative. Then a few nights later, their father had called, because Mrs. Santos was irrational and alarmed. "Our father thought Mother might have had a seizure," one daughter said.

In spite of Mrs. Santos's combativeness, I did a brief neurological exam. Then holding her small hands in mine, I noticed her unpolished nail beds and the fine white lines stretched across them as if they had been painted on, but they were not. I immediately ordered an analysis of her nails, hair, and urine and started safe and effective treatment. Had I not taken her symptoms seriously and searched for other clues in the examination, she may have been given strong mind medications (pharmaceuticals) that may have masked the symptoms without treating the toxicity.

I tell my medical students at USC there are three things a doctor must be certain of before starting treatment: *a probable diagnosis, a proper diagnosis, and a treatable diagnosis.*

You may wonder what this has to do with the medical treatment of Alzheimer's disease. Remember, the commencement of medical treatment is always dependent on the diagnosis. Did Mrs. Santos have Alzheimer's disease? Was it another type of dementia or mental illness? Why would I analyze her nails and hair?

Turns out, Mrs. Santos suffered from delirium because of arsenic poisoning. Arsenic causes white lines in the nail beds, which I noticed when I held her hands during the examination. Was someone trying to poison Mrs. Santos? Not at all. Her daughter said that relatives in Mexico had sent the woman a case of jam made from apricots and apricot pits to help prevent the spread of her cancer. In addition, the jam had high levels of arsenic—that I determined when noticing the common signs of white lines in her nail beds. It makes sense, in that apricot pits contain traces of arsenic. Evidently, by eating the apricot jam, enough arsenic built up in Mrs. Santos's system to cause symptoms that mimicked a brain disorder or Alzheimer's disease.

As you've witnessed in most of my patient narratives, in my clinical experience both men and women are generally brought to see an Alzheimer's specialist by their children. I have seen, as have my colleagues, a tremendous amount of denial in marriages, that there is something "potentially destructive" going on in their spouse's brain. The healthy spouse does not want to admit that the "love of their life" has a problem. When I asked Mrs. Santos's husband about her mental state and memory loss, he said he didn't notice his wife's severe mental problems, saying, "Well, she has her bad moments, but I think she'll snap right back, Doctor."

Usually, the wives, in general, tend to be more realistic. Much has to do with their level of sophistication and experience about Alzheimer's disease, perhaps from knowing someone with Alzheimer's or caregiving to a family member with the same.

Still, adult children are the ones most sensitive to change—as you noted with Mrs. Santos being brought to the ER by her two daughters. Spouses seem to miss the gradual deterioration of their partner's mind. Most commonly, a calamity or potential crisis is the cause of a brain evaluation. This may stem from a car accident, nighttime confusion as the patient is found wandering through the neighborhood at 3:00 A.M., or other precipitating event.

In this last chapter, I want to help you understand several important concepts regarding the medications and procedures available for the treatment of Alzheimer's. In doing so, I will answer the following questions:

1. When is medication necessary for Alzheimer's disease?
2. When is medication considered to be a drug as compared to a natural dietary supplement?
3. How can medications help *and* harm you?
4. What's the honest truth about the efficacy of Alzheimer's medications versus the financially driven hype?
5. What other medications are used in the treatment of symptoms of Alzheimer's disease?
6. What new treatments are in the pipeline . . . for the future?

ENHANCING THE MIND

The past century has brought myriad false promises in the arena of mind-enhancing products. For example, at the turn of the last century, heroin and cocaine were commonplace—both given as treatment to ill patients and even taken by prominent physicians such as German psychiatrist Sigmund Freud. In the 1960s, the hallucinogen LSD—promoted by such notables as Harvard psychiatrist Dr. Timothy Leary—was thought to be the new mind-enhancing wonder drug, until numerous psychoses including suicides occurred that were directly related to this psychedelic drug.

Sure, we all want a cure for Alzheimer's disease. The more we see Alzheimer's rob our family and friends of their minds and memories, the more passionate we become in seeking effective treatment. To that end, there will always be some manic Medicine Men with their wagons ready to separate your wallet from your good common sense—from what you know to be true: *There is no magic pill to cure Alzheimer's.*

If you or a loved one has Alzheimer's, it's important to realize that while there are a few proven medications that can give temporary relief from this devastating illness, no medication is proven (yet) to fully manage this brain disease effectively. Also keep in mind that a well-known pharmaceutical company in one of its recent newsletters stated the truth: "All the treatments so far (including anticholinesterase inhibitors such as Aricept) have not reduced patients from being placed in nursing homes by even one day."

So, is that important? Absolutely. The American Academy of Neurology (AAN) notes that by the time the average patient is diagnosed with Alzheimer's disease, they still live another 8.5 years. On the average, they are admitted to a nursing home within two and a half years of the diagnosis. This means that if you get Alzheimer's, you might spend the last six years of your life without the company of your family or even the house that you called "home."

As we try to understand medical research, I believe it's vital that we remember four key points:

1. Scientific results that appear to be a gain in animal experiments may not translate into long-term help for the human mind.
2. Scientific trials must be controlled and then evaluated and retested over time to see if the signs and symptoms of Alzheimer's disease improve.
3. The time from concept of animal studies to an effective, safe, and proven drug is five to ten years at best.
4. The benefit of any pharmaceutical treatment must far outweigh the burden.

Natural Dietary Supplements and Home Remedies

The use of herbal or natural pharmacology as "home remedies" is increasingly commonplace as the cost of medicine soars. Yet, before you ingest any natural dietary supplements from your grocer's shelf, it's important to know the effects, the side effects, and how some of these "natural therapies" may have druglike consequences when taken with other prescribed medicines. As an example, the natural herbs ginkgo biloba and passionflower increase bleeding. When these herbs are taken with aspirin or warfarin, a prescribed anticoagulant, the chances for bleeding are even greater. In addition, when the botanical evening-primrose oil, available for just dollars at most groceries, is taken with the antidepressant Prozac, it can increase the likelihood of seizures.

Even though natural dietary supplements, including herbs, are not governed by the FDA and are distributed as "food products," they still have powerful, druglike influences on the body. Read about the potential effects they can have if you mix them with common medications. Talk to your doctor about your medications and supplements and see what changes need to be made.

Mind Medications Versus Holistic Remedies

Mind medicines are man-made drugs (pharmaceuticals) that affect the brain. Holistic remedies are naturally occurring substances in nature such as ginkgo biloba. When mind medications are ingested with some natural dietary supplements, they vie for the same transporter systems. Each medication and supplement has specific actions, interactions, and side effects. Caffeinated tea is one example of the good and bad found in holistic therapies. While green and black tea are proven to be excellent antioxidants that can fight free radicals in the body, the additional caffeine may produce anxiety and sleeplessness, and also interact with psychotropic medications. Caffeine, an addictive "drug," is often responsible for insomnia as well as withdrawal headaches and irritability.

When to Consider Mind-Altering Drugs

You may wonder at what point you should consider taking a mind-altering drug—whether a prescribed medication or holistic botanical or natural dietary supplement. With the risks involved, it makes sense to talk to your doctor before self-medicating with any medication or supplement that may alter your mind. Here are some guiding recommendations:

Mind medications. If you have experienced memory problems consistently for three months, then your primary care doctor may consider prescribing a mind medication (pharmaceutical). Nevertheless, before you panic and think you have Alzheimer's disease, make sure your physician has ruled out other treatable causes of memory loss such as thyroid disease, depression, vitamin B_{12} deficiency, and recent gastric bypass surgery among other problems discussed in chapter 8.

Also, I emphatically advise you to consult with a specialist (a board-certified neurologist), especially if the medication is to be continued for more than two months. *Most mind-enhancing drugs are proven ineffective.* In addition, mind-enhancing drugs are extremely costly and even dangerous. On the side, if you've read the latest media recommendations to reverse Alzheimer's disease, along with the prime-time advertising by major pharmaceutical companies, you'd come up with a curious concoction of ginkgo biloba, Hydergine, Prozac, and Aricept. The problem is that *not one* of these natural dietary supplements and mind medications has been scientifically proven to increase the longevity of brain function.

Holistic or natural supplements. Holistic supplements are over-the-counter natural dietary supplements such as ginkgo biloba, kava, Saint-John's-wort, and valerian, among others. Consultation with your health care professional is always advised before taking a holistic drug. Medications such as aspirin interact with these non-FDA-approved natural dietary supplements, and just because they are sold at the corner supermarket does not mean they are safe—or effective. Complications occur frequently in older adults who mix various medications with supplements.

Treating the Symptoms of Alzheimer's Disease

In treating Alzheimer's disease today, the best we can do is treat the symptoms—since there is no cure, nothing to halt and reverse the disease once it begins. According to the American Academy of Neurology, the major complications of Alzheimer's disease that are of most concern to the family or caretaker include the following signs and symptoms:

1. Aggressive behavior
2. Illusions, hallucinations, and psychosis
3. Bowel and bladder incontinence
4. Insomnia and wandering

Aggressive Behavior

Aggressive behavior is found in 65 percent of all patients with Alzheimer's disease. This aggression is most commonly at night and occurs frequently during the transition from mild to moderate Alzheimer's. Medications that are best to deal with it are the neuroleptics such as Seroquel, Zyprexa, and Geodon that have low side-effect profiles. Other antipsychotics such as Haldol, Thorazine, and Risperdal cause Parkinsonian symptoms, such as rigidity and sedation, and increase the chance of falls and injury.

These medications, according to recent advice from the *New England Journal of Medicine*, must be closely monitored. It is advised that a specialist monitor these medications, as long-term treatment often is not necessary. That is, once a patient is aggressive, it does not mean he will continue to be so.

Illusions, Hallucinations, and Psychosis

Hallucinations, illusions, and psychosis are behavior symptoms that often cause great distress to the family, especially to the spouse or a child. During this time, the Alzheimer's patient begins to develop paranoid behavior. Since they cannot remember or grasp what is happening and become overwhelmed easily, they begin to imagine things. A similar experience occurs when people first lose their hearing. They begin to think people are whispering about them, believing that others are trying to keep secrets from them. In Alzheimer's disease, patients cannot remember where they placed their wallet, keys, or cell phone. They begin to believe that people around them are plotting against them, taking these items and deliberately hiding them. The Alzheimer's patient may become accusatory, most often to those they loved and trusted the most. This can then lead to more aggressive behavior.

Alzheimer's patients often get illusions first. This illusion is the misidentification of a real object. They might see a coat hanging on a coatrack and believe it's a person—that an intruder is in the house. An Alzheimer's patient might see an old photograph of her mother and then tell other family members the next morning that their mother had come to visit them during the night. Yet when the photo is removed, the illusions go away. Alzheimer's patients imagine people have come to visit them and they might vividly describe the people and the visit. They'll even tell you stories about someone who came and told them to do a task or make a move.

Delusions and psychosis, fixed false beliefs that are persistent, are, thankfully, uncommon in Alzheimer's disease. The hallmark of psychosis is that it is a false, fabricated reality—but a reality the patient believes is true and nothing can change his or her mind. For example, one of my patients shared his story that a plane flew overhead to X-ray his house to determine if the patient was doing something wrong. Each time I saw the elderly man, he'd relay the same story with greater detail than before. Despite being told that there were no airports or overflying airplanes, nor was there any reason for someone to spy on him, he persisted in his belief and only became angry and accused his own wife of being part of the conspiracy. In Alzheimer's disease, such fixed, false beliefs are not common. If there is a delusion in Alzheimer's, it is often random. This particular patient had a long history of schizophrenia.

Delusions must be distinguished from confabulations. A confabulation is a "fill in the gap" response. It's what neurologists used to call the "string sign," that is, when a person with limited comprehension abilities makes up a plausible story. For instance, let's say a physician raises both hands and asks, "What color string am I holding?" A normal person would realize that there wasn't any string in the doctor's hands. However, a person who confabulates will go on to describe the color, size, and length of the string. If a bill goes unpaid, the person that confabulates says that it was sent to the wrong address. Or looking at milk spilled on the floor, this person would say the cat did it by opening the refrigerator and pushing it over. Or the person might say there's a leak in the pavement, and it just "looks" like milk. Sometimes the confabulation can be quite inventive.

Here's some advice: Never, ever confront the patient who confabulates; never call them liars, because it causes agitation and resistance. Instead, listen attentively. If they have not endangered themselves, such as by leaving the stove on high, simply clean up the problem (the spilled milk) and say that it's okay to spill milk. It just needs to be wiped up. The same medications used to treat aggression can be used for these patients, especially if the behavior, hallucinations, or illusions are frightening and cause distress to the patient. A simple treatment may also include a night-light and placing familiar objects in the room.

Patients who experience hallucinations or illusions need twenty-four-hour supervision and should not be left home alone, especially if they demonstrate activities that might be a danger to themselves or to others, such as cutting electrical wires or leaving the stove on. Try to maintain a stable, predictable environment with assistants who are familiar with the patient's cultural background if possible.

Bowel and Bladder Incontinence

In the early and mid-stages of Alzheimer's disease, bowel and bladder incontinence is usually not a problem. But as the disease progresses to the mid- to late stages, bowel and bladder incontinence can occur. Management of this problem includes:

- Scheduling bathroom visits
- Wearing pads
- Using bladder inhibitors for urinary incontinence (these are often dangerous as they can cause urinary retention)

Insomnia and Wandering

The Alzheimer's patients' biorhythms are easily disturbed. These rhythms are worsened when they are allowed to sleep during the day, and their sleep-wake cycles become more disturbed than normal.

In 30 percent of Alzheimer's patients, sleep disturbance is a major reason they're placed in nursing homes, as the caregiver becomes exhausted. Sleeplessness at night is often associated with wandering about 20 percent of the time. The wandering patient leaves his home and walks the streets of his neighborhood. In their home or in a nursing home, they might get into the rooms or beds of other people living with them. It's completely common for an Alzheimer's patient to be found snoozing in a bed across the hall from her room or in another wing of the nursing home's Alzheimer's unit. You can imagine the havoc this causes to both the patient and the recipient!

Treatment involves managing sleep patterns by keeping the patient active during the day and having a fixed schedule. Sleep-onset drugs, such as Ambien, are a last resort, as are antipsychotics.

ALZHEIMER'S MEDICATIONS

When someone has Alzheimer's disease, nerve cells and vital chemicals in the brain are lost over time. This occurs in parts of the brain that are vital to memory and other mental processes. Let's look at several of the most common medications approved by the FDA for treatment of the symptoms associated with Alzheimer's disease. These medications work in specific ways to boost memory or slow down the progression of Alzheimer's.

Acetylcholinesterase Inhibitors

Acetylcholinesterase inhibitors, including Aricept, Exelon, and Razadyne/ER, reduce the destruction of acetylcholine after it's been excreted, thereby leaving it around for longer periods of time. The rationale behind the use of acetylcholinesterase inhibitors is that it has been noticed in Alzheimer's disease that there was a decrease in acetylcholine at the receptor site. Therefore, anything that would increase the amount of acetylcholine would improve memory.

While acetylcholinesterase inhibitors may temporarily improve the patient's symptoms, these medications do absolutely nothing to rectify the disease process. While definitely not a cure, there's some inclination that acetylcholinesterase inhibitors may delay the disease process.

Donepezil (Aricept)

How It Works: Aricept is a cholinesterase inhibitor that stops the breakdown of acetylcholine, a chemical in the brain used for memory and other mental functions. In Alzheimer's disease, there is a deficiency in acetlycholine in some areas of the brain, which accounts for some symptoms of the disease. Cholinesterase inhibitors also help increase the levels of acetylcholine in the brain. By increasing the amount of acetylcholine, it's thought that communication between cells should improve and thus increase memory.

The problem is that Alzheimer's causes a loss of the communication wires and neuron cells. Adding more transmitters (acetylcholine) will only provide temporary help. It does not fix the problem.

Possible Side Effects: Aricept has the fewest side effects of all the anticholinergic drugs. About 5 to 10 percent of patients may have mild side effects such as nausea, vomiting, and diarrhea. Some patients experience weight loss and insomnia.

Rivastigmine (Exelon)

How It Works: Exelon is thought to maximally decrease the progression of Alzheimer's disease, by blocking acetylcholine and butyrocholinesterase. Of all the medications for Alzheimer's disease so far, Exelon seems to be the only one that works in this way. This medication also affects butyrocholine, which is another type of cholinergic transmitter in the brain.

Possible Side Effects: With Exelon, side effects might include sweating, diarrhea, and nausea. Other less common side effects include seizures and arrhythmias. (It now comes in a patch [Exelon Patch] that is paced on the skin once a day. A marked reduction in side effects has been noted with the patch.)

Galantamine (Razadyne)

How It Works: Razadyne blocks acetylcholine but not butryocholin-esterase. Razadyne also binds to nicotine receptors, which increases neurotransmitters for brain activity. Razadyne may also block cell death in Alzheimer's disease, which might slow the progression of Alzheimer's.

Possible Side Effects: Side effects with Razadyne include upset stomach, vomiting, diarrhea, and urinary tract infections (UTIs). Other problems include arrhythmias and seizures.

Memantine HCL (Namenda)

How It Works: Namenda is approved by the FDA for treating moderate and severe Alzheimer's disease. Namenda works by blocking n-methyl D-aspartate (NMDA) receptors to reduce the entry of calcium into the neurons, which may protect it from damage. The NMDA receptor binds the neurotransmitter glutamate, thus increasing its charge and making it more likely to transmit its message to the next neuron. Namenda may also block programmed cell death. I have seen Namenda improve speech, learning, and recent memory in Alzheimer's patients, and I highly recommend this one.

Possible Side Effects: Some side effects associated with Namenda include dizziness, headache, confusion, Stevens-Johnson syndrome (a life-threatening allergic reaction), and seizures.

Antidepressant Medications

Depression is seen in over 70 percent of Alzheimer's patients and complicates the disease by interfering with sleep and energy. Antidepressants are commonly prescribed to alleviate the signs and symptoms of depression and, thus, improve the quality of the person's life in spite of the Alzheimer's disease.

Antidepressants are effective in the early stages of Alzheimer's when patients recognize their cognitive loss, and the loss of their mental abilities cause stress, with consequent increase in cortisol levels that eventually produces a decrease in serotonin, dopamine, and even adrenaline. Patients' first symptoms include difficulty with sleep, early morning awakening, increased fatigue, severe anxiety, and finally weight loss and anhedonia (a complete loss of pleasure). The patient can get to the point where she is immobile and appears to be severely demented. Severe depression can masquerade as Alzheimer's. I have had at least 100 cases in which severe endogenous depression was the primary cause of memory loss. SSRIs and SNRIs are the primary antidepressants used. There's also another category of antidepressants that contains bupropion (Wellbutrin), trazodone (Desyrel), and duloxetine (Cymbalta).

Tricyclics
Amitriptyline (Elavil)
Doxepin (Sinequan)
Nortriptyline (Pamelor)

The tricyclics were the first generation of antidepressants, including amitriptyline (Elavil), doxepin (Sinequan), and nortriptyline (Pamelor). These are drugs that reuptake serotonin, increasing the amount available for use. Serotonin is the calmative neurotransmitter that also enhances dopamine production (the feel-good transmitter). Initially, these drugs were to be used as antihistamines for allergies until it was found that people who took the tricyclics received a great antidepressant effect. Their sleeping habits improved and their energy levels increased, as did their appetite. Today, the tricyclics are used as both antidepressants and pain relieving modulators. In particular, amitriptyline is used for general neuropathic pain, migraines, arthralgias, and myalgias. Due to arrhythmia of the heart that they may cause they are not routinely recommended for patients over sixty.

Selective Serotonin Reuptake Inhibitors (SSRIs)
In the mid- to late 1980s, the second generation of antidepressants, the selective serotonin reuptake inhibitors, or SSRIs, was formulated. These started with Prozac and then moved to sertraline (Zoloft), paroxetine (Paxil), and fluvoxamine maleate (Luvox). Other antidepressants with different mechanisms from the SSRIs were also developed during that

same time, including bupropion (Wellbutrin), venlafaxine (Effexor), and mirtazapine (Remeron).

Antidepressants launched in the 1990s have completely different biochemical effects from the drugs of the past. This is important because we have made progress in diagnosing new subtypes of depression and have also arrived at new genetic findings, leading to the new field of pharmacogenetics where a given drug's metabolism and clinical effects may be genetically determined. Most of the new antidepressants, such as SSRIs, can be used with no dietary restrictions. They are safe even for cardiovascular patients, including those with cardiac arrhythmias and blood pressure changes.

As I stated in chapter 7, I believe that the selective serotonin reuptake inhibitors (SSRIs) may *decrease* the chance of Alzheimer's as they elevate serotonin levels in the brain while also suppressing dopaminergic pathways. Many patients who take antidepressants report more refreshing sleep and clarity of mind. Because they feel relaxed, in control, and rested, they can make healthy lifestyle choices and deal with life's challenges in a much healthier manner.

SOME COMMONLY PRESCRIBED ANTIDEPRESSANTS IN ALZHEIMER'S DISEASE

Generic Name	Brand Name	Common Usage
amitriptyline	Elavil	Sleep
doxepin	Sinequan	Sleep stage 4
escitalopram	Lexapro	Anti-anxiety
fluoxetine	Prozac	Mood enhancer
sertraline	Zoloft	Mood enhancer
paroxetine	Paxil	Mood enhancer
venlafaxine	Effexor	Anti-anxiety
duloxetine	Cymbalta	Anti-anxiety
trazodone	Desyrel	Sleep enhancer
bupropion	Wellbutrin	Mood enhancer

Neuroleptics, such as Haloperidol (Haldol)

In Alzheimer's patients, the neuroleptics are given to control serious psychological behavior such as combativeness, illusions, hallucinations, and delusions. Ideas of reference, such as when a patient believes that someone is out to get them, are also treated with neuroleptics.

How They Work: Haldol is a commonly used neuroleptic. Haldol is considered to be particularly effective in the management of hyperactivity, agitation, and mania and is used to treat acute and chronic psychosis, including schizophrenia and manic states. Haldol is also used in the management of aggressive and agitated behavior.

Possible Side Effects: Haldol can cause insomnia, headaches, cerebral seizures, and tardive dyskinesia, or involuntary movements of the lips, tongue, face, trunk, and extremities.

Atypical Neuroleptics, such as Quetiapine (Seroquel)

How They Work: The atypical neuroleptics are prescribed for psychosis, severe anxiety, and wandering. While Seroquel is not indicated for the treatment of dementia-related psychosis, it is used at low doses to offset behavioral/psychosis problems with Alzheimer's.

Possible Side Effects: Side effects of Seroquel include headache, somnolence, hypertriglyceridemia (high levels of triglycerides), severe hypotension, tardive dykenesia, and neuroleptic malignant syndrome (NMS), a life-threatening neurological disorder often caused by an adverse reaction to a neuroleptic or antipsychotic medication. (This is rare at low doses.)

Other Medications Used with Alzheimer's Disease

Olanzapine (Zyprexa)

How It Works: Olanzapine is a selective monoaminergic antagonist that's normally prescribed for schizophrenia, or manic episodes of bipolar disorder. It helps in controlling symptoms such as hallucinations and delusions, as well as social withdrawal and apathy.

Possible Side Effects: Some possible side effects of Zyprexa include dizziness, daytime sleepiness, weight gain, neuroleptic malignant syndrome, diabetes mellitus, and extrapyramidal symptoms such as tremors, rigidity, drooping, rolling eyes, and a masklike expression.

Benzodiazepines

Alprazolam (Xanax)	Oxazepam (Serax)
Diazepam (Valium)	Temazepam (Restoril)
Lorazepam (Ativan)	Triazolam (Halcion)

How They Work: The benzodiazepines belong to a group of medications called central nervous system (CNS) depressants. These medications act on neurotransmitters to slow down normal brain function. CNS depressants are commonly used to treat anxiety and sleep disorders. These drugs are all habit-forming or addictive.

With Alzheimer's disease, I avoid the use of benzodiazepines to ameliorate symptoms, even anxiety and agitation. I advise this for the following reasons:

1. Patients can develop a tolerance to daily use of benzodiazepines within a month.
2. Patients become addicted, so when the drug is withdrawn, they will have increased agitation.
3. Increased doses of benzodiazepines are required to maintain the same effect. With an increased dosage, the patient may have increased confusion and a deterioration of both recent and past memory. Functional MRIs have shown that the use of benzodiazepines is associated with an overall decrease in brain activity.

Possible Side Effects: Side effects of benzodiazepines may include memory impairment, psychomotor retardation, toxicity, depression, and emotional blunting. These drugs may also give rise to physiologic dependence.

Medications Used to Treat Insomnia

Initially, I may prescribe eszopiclone (Lunesta), zaleplon (Sonata), zolpidem (Ambien), or ramelteon (Rozerem) for minor sleep disturbance. If these don't work, I often recommend an antidepressant such as Sinequan, trazodone, or amitriptyline. Amitriptyline, due to its arrhythmic effects, is not advised in patients with coronary artery disease or over sixty years of age.

Medications to induce sleep should be monitored and discontinued when feasible. Unfortunately, in many nursing homes that I attend, the patients with dementias such as Alzheimer's often take many sedatives, antidepressants, and antipsychotic medications—without one doctor reviewing this list. As an example, one elderly man was on seventeen medications because of his wandering, aggressiveness, and psychotic behavior. The medications had accumulated in his system, causing liver failure and death. Make sure one doctor sees all the medications you take to avoid serious complications or toxicity.

SLEEP MEDICINES

Purported to be Non-Habit-Forming

Generic Name	Brand Name
eszopicione	Lunesta
zaleplon	Sonata
ramelteon	Rozerem

Non-Habit-Forming

mirtazapine	Remeron (atypical antipsychotic)
quetiapine fumarate	Serequel (atypical antipsychotic)
trazodone	Dyserel (antidepressant)

Future Treatment:
What's in the Pipeline for Alzheimer's Disease

The studies are ongoing in a desperate search for an Alzheimer's cure. Some studies are focusing strictly on the hope of an Alzheimer's vaccine, an immunization that can finally rid us of this horrific disease. Others are looking at nerve-growth factors, monoclonal antibodies, and stem cells—hoping for a modern miracle. Still others are focusing on the beta-amyloid hypothesis, because some believe that by blocking the generation of beta-amyloid in the brain or enhancing the clearance of beta-amyloid, they can successfully treat those with Alzheimer's disease. Let's look at some of the more promising theories in the search for an Alzheimer's cure.

Alzheimer's Vaccine

In 1999, it was announced to the medical community that injecting beta-amyloid into mice genetically predisposed to form amyloid plaque, and Alzheimer's disease, actually prevented the amyloid from being produced. In the young mice that were injected, no plaque was found after their maturation. However, in the older mice that already had plaque formed, no change was noted. The conclusion of this study? The young mice produced antibodies against beta-amyloid, thereby preventing the disease. In the human trials, this study has been abandoned because of the death of six human participants from encephalitis.

Growth Factor

The concept of growth factor is simple, logical, and obvious. We know that growth factors influence stem cells. These growth factors are like sentinels, watching their territory of the brain and making certain it works in harmony with the rest of the cells. While growth factors can influence all the cells in their domain, scientists are still determining how to manufacture growth factor, similar to the way we manufacture insulin to treat diabetes.

Monoclonal Antibody Treatment

Elan Pharmaceuticals is now approaching the treatment of Alzheimer's focusing on the beta-amyloid hypothesis. It is believed that by blocking the generation of beta-amyloid or enhancing the clearance, it will result in the successful treatment of Alzheimer's patients, again a logical hypothesis. Beta-amyloid is involved in the formation of plaque that causes difficulty in thinking and learning. The fundamental idea is that the clearance of beta-amyloid may lead to improved function in Alzheimer's patients. Beta-amyloid, also known as ABETA, is actually a small part of a larger protein called amyloid precursor protein or APP. Beta-amyloid is formed when certain enzymes called secretases cleave the APP. We are starting clinical trials at our clinic. See www.anti-alzheimers.com.

Beta-Amyloid Immunotherapy

Beta-amyloid immunotherapy treats amyloid disease by enhancing the body's immune response. Active immunization much like that in the polio vaccine stimulates the body's own immune system to produce anti-beta-amyloid antibodies that may attach to the beta-amyloid and clear it from the brain.

So far, this treatment has reduced *tau protein*, which is a known marker that's elevated in the CSF. Also, brain volume was lowered in anti-beta-amyloid responders as measured by MRI. It has not proven to be effective yet in humans.

Secretase Inhibitor Research

Beta-secretase is believed to initiate the first step in beta-amyloid formation, the precursor to plaque development in the brain. Gamma-secretase is a multiprotein complex thought to play a significant role in the formation of beta-amyloid. Gamma-secretase inhibitors appear to reduce beta-amyloid levels in the brain.

Recent results of myriad clinical trials conclude that researchers are unable to reduce the rate of Alzheimer's disease in men and women who've participated in the studies. There is no medical treatment in the near future, and anyone who says otherwise is giving false hope. *Prevention of Alzheimer's disease is the only reasonable alternative, as outlined in this book.*

Stem Cells

The most recently purported miracle cure for Alzheimer's disease (and many other nervous system diseases) is said to be the "stem cell." Stem cells are immature cells that could be matured into brain cells. One way stem cells could be used is to replace dead brain cells such as those cells that are lost because of Alzheimer's disease. Indeed, some stem cells have been used to cure diseases of the blood for more than thirty years, so this is an exciting concept. And although this is possible in theory, if a person has lost his memories because of Alzheimer's disease, and we use stem cells to give him or her back the capacity to remember, that person will not regain those old memories and will be a different person from then on. This is the reason that prevention of Alzheimer's disease is so extremely important—*We cannot regain memories; they are forever erased from our brains once Alzheimer's disease takes hold.*

I am a firm proponent of using stem cells to treat the appropriate types of diseases. The medical community has been using them to cure blood diseases such as leukemia for decades. Before that, the diagnosis of leukemia was a diagnosis of death. So we have come an astonishing distance—because of stem cells. But even after thirty or more years of actual clinical application of this type of stem cell therapy (and after all the basic science and animal experiments that preceded this), we can still treat only about 70 percent of the patients, and much of the time severe rejection problems occur. We still have a long way to go with the one form of stem cell therapy that we are already using. It's going to be a long time before new therapies, particularly those for the brain, come into widespread clinical practice.

PREPARE NOW FOR YOUR FUTURE

Consider the following sage advice with the FIVE method:

F—Have a strong social and spiritual network that includes Family, Friends, and Faith. A recent study by AARP notes 94 percent of baby boomers believe there is a God, yet statistics show only 33 percent actively practice this belief in a faith community; 30 percent have no nearby family or close friends.

I—Insurance. Take care of your future financial needs with elder care Insurance.

V—Vest in a retirement community that has step-down care. Baby boomers are searching for retirement homes in remote areas in the mountains, valleys, and the far west, even in foreign countries. Many of these foreign countries do not have the money or the inclination to treat people over age sixty, especially if they're foreigners.

E—Evaluate who will be on your health care team if and when you do have Alzheimer's disease. Prepare a team including a physician in geriatric care young enough to be around when you will need this health care expert. Have a hospital with a good reputation that is within thirty minutes of where you live. Be certain that there is a rehabilitation unit that takes your insurance so you are not relegated to a nursing home.

AND THE GREATEST OF THESE IS HOPE

Though science has made significant gains in the struggle with Alzheimer's, there's still no sure cure for this disease. However, medical studies have clearly indicated that there is hope through *prevention*. We must all remember that the greatest resource that God has given us is our body, mind, and soul. Through the four steps, I've explained how to maximize brain health and, more importantly, how to maintain your motivation through balancing your hormonal symphony, putting you in concert with the natural means at your disposal to mitigate the aging process and mental deterioration.

Longevity with continued independence is completely possible today. All it takes is a decision to *act right now*. I have given you the tools in the 4-Step Anti-Alzheimer's Prescription. Now you must make these a daily priority.

A cure for Alzheimer's disease may come tomorrow or in fifty years.

However, prevention must start today. That said, I wish I could have helped my own father and my Italian relatives who died with Parkinson's and Alzheimer's disease. I urge you to join me and my wife, our family members and dear friends, my patients and their families, and *start the four-step program today!*

APPENDIX A

Anti-Alzheimer's Diet Shopping List
(*INCLUDING APPROVED BRAND NAMES*)

Nuts: Choose raw nuts, no salt or oils added, natural peanut butter, or other natural nut butters.

- Nuts and nut butters are healthy *and* calorically dense; watch the portions you eat, for they can add pounds if you overdo it.

Fruits: Açaí, apples, apricots, avocados, bananas, bilberries, blackberries, blueberries, cantaloupe, cherries, cranberries, grapes (red), kiwis, lemons, limes, mangos, melon, oranges, pineapple, peaches, pears, plums, raspberries, strawberries, tomatoes, unsweetened applesauce.

- Must have at least one serving of blueberries or other dark purple berry daily (¾ cup).
- Buy fresh local fruit in season or frozen fruits with no sugar added.

Vegetables: Artichokes, arugula, asparagus, broccoli, broccoli rabe, Brussels sprouts, cabbage, carrots, cauliflower, celery, cucumbers, eggplant, garlic, green beans, haricot beans, greens (turnip, mustard, collard), greens (mixed-variety lettuce), kale, onions, peppers (green, red, orange, and yellow), snow peas, spinach, squash varieties (yellow, acorn, and spaghetti), sweet potatoes.

- Dr. Praeger's veggie patties (California burgers, spinach and broccoli patties) are a great choice for the whole family.

- Purchase fresh local vegetables when possible or frozen varieties with no sauces or added salt.

Low-Fat Dairy: 1% or less milk, low-fat plain yogurt, low-fat cottage cheese (calcium fortified), low-fat cheese (Cabot 50–75% light).
- For children and pregnant and lactating women, organic products are recommended.

Soy Protein Foods: Boca Burgers, Dr. Praeger's California Veggie Burgers, low-fat plain soy milk, tempeh (soybean based, White-Wave brand), tofu (silken for smoothies, firm for stir-fries), soy deli meats (Yves brand), soy cheese (Veggie Slices by Galaxy Nutritional Foods).
- Keep in mind that many soy foods can be processed and high in sodium. If sodium levels are a concern, be sure to account for the sodium found in these products.

Protein Foods: White meat chicken and turkey with no skin, wild salmon, halibut, sardines, lean pork, Chilean sea bass, flounder, fluke, bluefish, pork tenderloin, eggs that contain omega-3 fatty acids, lean luncheon meats (turkey, ham; try Boar's Head, Applegate Farms).
- Children, pregnant and nursing women, and women planning families should avoid fish high in mercury, including shark, tilefish, swordfish, and king mackerel.

Legumes: Choose a variety of beans—black beans, chickpeas, great northern beans, kidney beans, lentils, lima beans, peas (dried), pink beans, red beans, soybeans (dried), white beans.
- Soak and boil dry beans.
- Purchase canned beans with no salt added and always rinse thoroughly.
- If you purchase canned beans that contain salt, rinse thoroughly with water to remove excess sodium.

Starches: Whole grains, such as brown rice, whole wheat couscous and pasta (De Cecco, DeBoles, Eden Organic), rye, barley, quinoa, oats, amaranth, millet, spelt, wheat berries, wild rice.
- Make sure packages say whole grain or whole wheat.
- If the main flour listed on the label is wheat or unbleached wheat flour, the product is not whole grain.
- Most multigrain, rye, oat, and pumpernickel breads in the United States are not whole grain.
- Breakfast foods: Flax waffles (Van's), Cheerios, Wheaties, Total, Shredded Wheat, oatmeal, whole wheat pancakes (Hodgson Mill mixes)

Dark Chocolate: 70% cocoa
- Combine chocolate with a serving of nuts (6 almonds or cashews, 2 whole walnuts, 10 peanuts).

Tea: Chamomile, green tea, and black tea.
- Please note: adding milk to tea may blunt the beneficial antioxidant effect of tea.

Red Wine or Grape Juice:
- Limit wine to one 4-ounce serving per day
- A serving of fruit can be exchanged for a 4-ounce glass of 100 percent grape juice. Examples of fruit servings: ¾ cup berries (raspberries, blueberries, blackberries), 1¼ cups whole strawberries, 1 medium-size apple, peach, orange, ½ grapefruit, 4-inch banana, 10 grapes, 2 tablespoons raisins, ½ cup fruit salad, ½ mango, ½ cantaloupe.

Spices and Condiments: Turmeric powder, dried rosemary, garlic powder, fresh black peppercorns, sea salt, balsamic vinegar, cider vinegar, rice vinegar, cold-pressed olive oil.

Treats: Treats should be limited in the diet. However, when you feel the need to have a "sweet treat," it should be no more than 150 calories (figure out calorie levels by reading the Nutrition Facts label).

- Examples include: Newman's Own cookies, sorbet, low-fat frozen yogurt, rice/chocolate/vanilla pudding (Kozy Shack).
- Purchase prepackaged foods to help contain calorie limits such as 100-calorie packs, Skinny Cow ice cream sandwiches, Kozy Shack pudding snacks.

DAY 1

2,000 kcal

BREAKFAST

- 100% whole wheat English muffin
- 4 teaspoons natural peanut butter
- ½ banana
- 1 cup skim milk

LUNCH

- 5 ounces chicken salad made with 1 tablespoon of light mayo and turmeric
- 2 slices light whole wheat bread
- ½ cup baby carrots
- 4 ounces tomato juice

DINNER

- 8 ounces salmon
- 1 cup broccoli, steamed with garlic cloves
- ⅔ cup whole wheat couscous
- 4 ounces red wine or grape juice

SNACK 1

- 1 cup nonfat plain yogurt with ¾ cup blueberries and 1½ teaspoons sunflower seeds

SNACK 2

- ¾ cup low-fat cottage cheese with ¾ cup blackberries, and cinnamon for flavor

DAY 2

2,000 kcal

BREAKFAST

- 2 eggs with ½ cup asparagus and 1 ounce of low-fat cheese
- 1 slice whole wheat bread
- 1¼ cups whole strawberries
- 1 cup non- or low-fat plain soy milk

LUNCH

- ¾ cup tuna fish made with 1 tablespoon light mayo
- Place tuna over 1 cup baby spinach
- 10 whole grain, low-fat crackers

DINNER

- 7 ounces grilled chicken
- 1 cup roasted Brussels sprouts
- ⅔ cup brown rice
- 4 ounces red wine or grape juice

SNACK 1

- 1 apple
- 4 teaspoons natural peanut butter

SNACK 2

- ¾ cup low-fat cottage cheese
- ¾ cup blueberries
- 6 almonds

DAY 3

2,000 kcal

BREAKFAST

- ½ cup oatmeal, cooked (cook with 1 cup of skim milk)
- ¾ cup blueberries
- 1 tablespoon crushed walnuts
- 2 hard-boiled eggs

LUNCH

- 5 ounces turkey
- 1 ounce low-fat cheese (Alpine Lace)
- 2 slices light whole wheat bread
- Mustard
- 1 cup non- or low-fat plain soy milk

DINNER

- 8 ounces shrimp
- ½ cup sautéed vegetables (spinach, tomato, mushrooms, broccoli, garlic, etc.) cooked with 2 teaspoons olive oil
- 2 tablespoons Parmesan cheese
- 1 cup whole wheat pasta, cooked

SNACK 1

- 1 ounce dark chocolate, 70% cocoa
- 12 cherries

SNACK 2

- 1 cup low-fat cottage cheese
- ¾ cup blackberries

DAY 4

2,000 kcal

BREAKFAST

- ¾ cup whole grain, unsweetened cereal
- ¾ cup blueberries
- 1 cup skim milk

LUNCH

- 2 slices rye bread
- 6 slices veggie ham
- 1 ounce Swiss cheese
- Mustard
- ½ cup baby carrots

DINNER

- 8 ounces breaded chicken cutlet (¼ cup bread crumbs and 2 teaspoons olive oil)
- ⅓ cup whole wheat couscous
- 1½ cups kale, cooked with garlic

SNACK 1

- 1 apple
- 4 teaspoons natural peanut butter
- 1 cup non- or low-fat plain soy milk

SNACK 2

- ½ cup low-fat cottage cheese
- ¾ cup raspberries
- Cinnamon for flavor

DAY 5

2,000 kcal

BREAKFAST

- 100% whole wheat English muffin
- 4 teaspoons natural peanut butter
- ½ banana
- 4 ounces tomato juice

LUNCH

- 5 whole grain crackers
- 1 cup arugula and 1 cup veggies
- 4 ounces grilled chicken
- 1 ounce cheese, 75% light
- 1 hard-boiled egg
- 2 teaspoons olive oil with vinegar and mustard dressing

DINNER

- 8-ounce tuna steak with soy ginger marinade, and 1½ teaspoons sesame seeds
- ⅓ cup brown rice with ⅔ cup kidney beans
- 1 cup broccoli, steamed

SNACK 1

- 1 cup nonfat plain yogurt
- ¾ cup raspberries
- 6 almonds

SNACK 2

- 1 apple
- 1 ounce cheese, 75% light

DAY 6

2,000 kcal

BREAKFAST

- 1 slice whole grain toast
- 3 egg whites and 1 egg, scrambled
- 1 ounce cheese, 75% light
- 1¼ cups whole strawberries
- 1 cup non- or low-fat plain soy milk

LUNCH

- 2 slices whole grain bread
- 1 soy-based veggie burger
- 1 teaspoon light mayo and mustard, lettuce and tomato
- 4 ounces tomato juice

DINNER

- 8 ounces lean pork with *soy marinade
- ½ cup pasta with ⅔ cup beans
- 1½ cups cooked veggies with garlic
- 1 teaspoon olive oil in cooking

SNACK 1

- 1 ounce dark chocolate, 70% cocoa
- 12 cherries
- 1 cup skim milk

SNACK 2

- ½ cup low-fat cottage cheese
- ¾ cup blueberries

DAY 7

2,000 kcal

BREAKFAST
- 2 whole grain, low-fat waffles
- ½ banana, sliced
- 2 ounces Canadian bacon
- 1 cup skim milk

LUNCH
- ½ cup whole wheat pasta
- 12 ounces scallops
- 1 cup spinach, cooked
- 1 tablespoon pine nuts
- 1 teaspoon olive oil

DINNER
- 6-ounce 98% lean turkey burger (mixed with 1 egg white, garlic powder, 1 teaspoon Worcestershire sauce, pepper)
- 1 whole grain English muffin
- 1 ounce cheese, 75% light
- 1 cup carrots roasted with 2 cloves garlic, 1 teaspoon turmeric, and 1 teaspoon olive oil

SNACK 1
- 1 apple
- 4 teaspoons natural peanut butter
- 1 cup non- or low-fat plain soy milk

SNACK 2
- ¾ cup blueberries
- 1 ounce cheese, 75% light
- 3 cashews

DAY 8

2,000 kcal

BREAKFAST
- 1 egg + 3 egg whites and 1 ounce 75% light cheese mixed with:
- baby spinach, tomato, onion, and ⅛ avocado
- 1 slice whole wheat bread
- ¾ cup raspberries in 1 cup nonfat plain yogurt

LUNCH
- 4 ounces turkey, low sodium
- 1 ounce Alpine Lace Swiss cheese
- 1 whole grain roll
- 2 cups dark greens and mixed veggies with 2 teaspoons olive oil, vinegar (to taste), pepper (to taste), and a dash of sea salt

DINNER
- *8 ounces cod
- 1 teaspoon olive oil in cooking
- Cooked with 2 cups veggies (baby spinach, onions, mushrooms, tomato), ⅛ cup dry white wine, 1 teaspoon turmeric; fresh lemon squeezed over fish
- Place over ⅓ cup brown rice with ⅓ cup lentils, cooked

SNACK 1
- *Brain-Boosting smoothie
- 1 cup skim milk
- 1 scoop soy protein powder
- ¾ cup blueberries, frozen

SNACK 2
- 1 cup fat-free cottage cheese
- ¾ cup pineapple

DAY 9

2,000 kcal

BREAKFAST
- ½ cup steel-cut oatmeal, cooked with:
- 1 cup skim milk
- 1 tablespoon crushed walnuts
- 2 hard-boiled eggs

LUNCH
- Sandwich made with 2 slices light whole grain bread and mustard
- 8 slices soy turkey
- 1 slice low-fat soy cheese
- 1 cup baby carrots, marinated in ½ cup balsamic vinegar and 1 teaspoon turmeric

DINNER
- Appetizer: 5 whole grain crackers and 2 ounces 75% light cheese
- 8 ounces salmon, baked or broiled
- 1 cup cooked kale made with 1 teaspoon sesame oil, pepper, and fresh garlic
- 6-ounce baked yam

SNACK 1
- *Brain-Boosting Peanut Butter Smoothie
- ½ cup nonfat plain yogurt
- ½ banana
- 2 teaspoons natural peanut butter
- 3 ice cubes

SNACK 2
- ¾ ounce freeze-dried blueberries
- 6 almonds

DAY 10

2,000 kcal

BREAKFAST
- 100% whole wheat English muffin
- 4 teaspoons natural peanut butter
- ½ banana
- 1 cup non- or low-fat plain soy milk

LUNCH
- 4 ounces chicken salad made with 1½ teaspoons light mayo and turmeric to taste
- 2 slices light whole wheat bread
- ½ cup baby carrots
- 4 ounces tomato juice

DINNER
- 8 ounces Chilean sea bass
- 2 cups spinach cooked with 1 teaspoon olive oil and garlic
- ⅔ cup quinoa
- 4 ounces red wine or grape juice

SNACK 1
- 1 cup nonfat plain yogurt with ¾ cup blueberries

SNACK 2
- 1 cup low-fat cottage cheese with ¾ cup blackberries and cinnamon for flavor

DAY 11

2,000 kcal

BREAKFAST
- 2 eggs with ½ cup tomato and onions
- 1 ounce low-fat cheese, 75% light
- 1 slice whole wheat bread
- ¾ cup blueberries
- 1 cup skim milk

LUNCH
- ¾ cup tuna fish made with 1½ teaspoons light mayo and turmeric to taste
- Place tuna over 1 cup mixed greens
- 10 whole grain, low-fat crackers

DINNER
- 7 ounces grilled chicken
- 1 cup roasted Brussels sprouts
- ⅔ cup brown rice
- 1 teaspoon olive oil in cooking

SNACK 1
- 1 apple
- 4 teaspoons natural peanut butter
- 1 cup non- or low-fat plain soy milk

SNACK 2
- ¾ cup low-fat cottage cheese
- ¾ cup raspberries
- 3 almonds

DAY 12

2,000 kcal

BREAKFAST
- ½ cup oatmeal, cooked (cook with 1 cup skim milk)
- ¾ cup blueberries
- 1 tablespoon crushed walnuts
- 2 hard-boiled eggs

LUNCH
- 5 ounces turkey
- 1 ounce low-fat cheese (Alpine Lace)
- 2 slices light whole wheat bread
- Mustard
- 1 cup non- or low-fat plain soy milk

DINNER
- 8 ounces shrimp, grilled
- ½ cup sautéed vegetables (spinach, tomato, mushrooms, broccoli, garlic, etc.) cooked with 2 teaspoons olive oil
- 2 tablespoons Parmesan cheese
- 1 cup whole wheat pasta, cooked

SNACK 1
- 1 ounce dark chocolate, 70% cocoa
- 12 cherries

SNACK 2
- 1 cup low-fat cottage cheese
- ¾ cup blackberries

DAY 13

2,000 kcal

BREAKFAST

- ¾ cup whole grain, unsweetened cereal
- ¾ cup blueberries
- 1 cup skim milk
- 1½ tablespoons flaxseed

LUNCH

- 2 slices rye bread
- 5 ounces turkey
- 1 ounce Swiss cheese, low-fat
- 2 teaspoons mustard
- ½ cup baby carrots

DINNER

- 12 ounces breaded flounder cutlet (¼ cup bread crumbs and 1 teaspoon olive oil)
- ⅔ cup kidney beans
- 1½ cups kale, sautéed with 2 garlic cloves, crushed

SNACK 1

- 1 apple
- 4 teaspoons natural peanut butter
- 1 cup non- or low-fat plain soy milk

SNACK 2

- ¾ cup low-fat cottage cheese
- ¾ cup raspberries
- 6 almonds and cinnamon for flavor

DAY 14

2,000 kcal

BREAKFAST

- 100% whole wheat English muffin
- 4 teaspoons natural peanut butter
- ½ banana
- 4 ounces tomato juice

LUNCH

- 5 whole grain crackers
- 1 cup arugula and 1 cup mixed vegetables
- 8 ounces flavored tofu (firm)
- 1 ounce cheese, low-fat
- 1 hard-boiled egg
- 1 teaspoon olive oil with vinegar and mustard dressing

DINNER

- 8-ounce tuna steak with soy ginger marinade, and 1 table-spoon sesame seeds
- ⅓ cup barley with ⅔ cup kidney beans
- 1 cup broccoli, steamed
- 1 cup non- or low-fat plain soy milk

SNACK 1

- 1 cup nonfat plain yogurt
- ¾ cup raspberries
- 6 almonds

SNACK 2

- 1 apple
- 2 ounces cheese, 75% light

DAY 15

2,000 kcal

BREAKFAST

- 1 slice whole grain toast
- 3 egg whites and 1 egg, scrambled
- 1 ounce cheese, 75% light
- 1 cup spinach
- 1¼ cups whole strawberries
- 1 cup skim milk

LUNCH

- 2 slices whole grain bread
- 1 soy-based veggie burger
- Mustard, lettuce, and tomato
- 4 ounces tomato juice

DINNER

- 6 ounces grilled chicken
- ½ cup pasta with ⅔ cup beans
- 1½ cups cooked veggies with garlic
- 2 teaspoons olive oil in cooking

SNACK 1

- 1 ounce dark chocolate, 70% cocoa
- 12 cherries
- 1 cup non- or low-fat plain soy milk

SNACK 2

- 1 cup low-fat cottage cheese
- ¾ cup pineapple

DAY 16

2,000 kcal

BREAKFAST

- 2 flax waffles
- 1¼ cups whole strawberries
- 3 egg whites + 1 egg with 1½ ounces cheddar cheese, 75% light
- 1 cup skim milk

LUNCH

- 1 cup tuna fish made with 1 tablespoon light canola mayo and turmeric
- Over 1 cup raw baby spinach
- 1 slice whole wheat toast

DINNER

- *4 ounces grilled tempeh, cooked with 1 teaspoon sesame oil
- 1½ cups steamed broccoli, cut into small pieces
- ⅔ cup black beans
- ½ cup barley
- Mix broccoli, beans, and barley, and add rice vinegar, sea salt, and fresh pepper to taste

SNACK 1

- 6 almonds
- 1 cup nonfat plain yogurt
- ½ banana
- Add cinnamon to taste

SNACK 2

- 1 cup fat-free cottage cheese
- ¾ cup fresh or frozen blueberries

DAY 17

2,000 kcal

BREAKFAST

- 2 eggs with ½ cup asparagus and 1 ounce low-fat cheese
- 1 slice whole wheat bread
- 1¼ cups whole strawberries
- 1 cup non- or low-fat plain soy milk

LUNCH

- ¾ cup tuna fish made with 1 tablespoon light mayo
- Place tuna over 1 cup baby spinach
- 10 whole grain, low-fat crackers

DINNER

- 7 ounces grilled chicken
- 1 cup Brussels sprouts, roasted
- ⅔ cup brown rice
- 4 ounces red wine or grape juice

SNACK 1

- 1 apple
- 4 teaspoons natural peanut butter

SNACK 2

- ¾ cup low-fat cottage cheese
- ¾ cup blueberries
- 6 almonds

DAY 18

2,000 kcal

BREAKFAST

- 4 4-inch whole wheat pancakes
- Pancake topping: ½ banana, mashed with 1 tablespoon crushed walnuts
- 1 hard-boiled egg
- 1 cup non- or low-fat plain soy milk

LUNCH

- 8-ounce tuna steak, grilled with soy marinade (1 teaspoon olive oil, 2 tablespoons soy sauce, 1 teaspoon turmeric)
- ⅓ cup whole wheat couscous
- 1 cup or 8 spears asparagus

DINNER

- 1 cup whole wheat pasta
- 4 soy meatballs, about 1 inch diameter
- 1 cup low-fat marinara sauce
- 4 ounces red wine or grape juice

SNACK 1

- 3 cups nonfat popcorn
- ¾ ounce freeze-dried blueberries
- 1 ounce cheese, 75% light
- 1 cup skim milk

SNACK 2

- 1 cup Greek yogurt
- ¾ cup blackberries
- 1 teaspoon vanilla

DAY 19

2,000 kcal

BREAKFAST

- 2 flax waffles
- ¾ cup blackberries
- 3 egg whites with 1½ ounces 75% light cheddar cheese
- 1½ cups non-fat plain soy milk

LUNCH

- *6 ounces grilled chicken salad, made with 1½ teaspoons light canola oil mayo and turmeric
- 1 slice whole wheat toast
- ½ cup raw string beans

DINNER

- *Tofu stir-fry (6 ounces firm tofu), cooked with 1 teaspoon olive oil
- ¼ pound shrimp
- 3 cups vegetables
- ⅔ cup brown rice
- ⅔ cup white beans
- 4 ounces red wine

SNACK 1

- 4 teaspoons natural peanut butter
- 1 apple

SNACK 2

- 1 cup fat-free Greek yogurt
- ¾ ounce freeze-dried blueberries

DAY 20

2,000 kcal

BREAKFAST

- 100% whole wheat English muffin
- 4 teaspoons natural peanut butter
- ½ banana
- 4 ounces tomato juice

LUNCH

- 5 whole grain crackers
- 1 cup arugula and 1 cup veggies
- 4 ounces grilled chicken
- 1 ounce cheese, 75% light
- 1 hard-boiled egg
- 2 teaspoons olive oil with vinegar and mustard dressing

DINNER

- 8-ounce tuna steak with soy ginger marinade, and 1½ teaspoons sesame seeds
- ⅓ cup brown rice with ⅔ cup kidney beans
- 1 cup broccoli, steamed

SNACK 1

- 1 cup nonfat plain yogurt
- ¾ cup raspberries
- 6 almonds

SNACK 2

- 1 apple
- 1 ounce cheese, 75% light

DAY 21

2,000 kcal

BREAKFAST

- 1 egg + 3 egg whites and 1 ounce 75% light cheese mixed with:
- Baby spinach, tomato, onion, and ⅛ avocado
- 1 slice whole wheat bread
- 1¼ cups whole strawberries in 1 cup nonfat plain yogurt

LUNCH

- 4 ounces turkey, low-sodium
- 1 ounce Alpine Lace Swiss cheese
- 1 whole grain roll
- 2 cups dark greens and mixed veggies with 1 teaspoon olive oil, vinegar to taste, pepper to taste, and a dash of sea salt

DINNER

- 8 ounces cod, baked or broiled
- 1 teaspoon olive oil in cooking
- Cooked with 2 cups veggies (baby spinach, onions, mushrooms, tomato), ⅛ cup dry white wine, 1 teaspoon turmeric, fresh lemon squeezed over fish
- Place over ⅓ cup brown rice with ⅓ cup lentils, cooked

SNACK 1

- *Brain-boosting smoothie
- 1 cup skim milk
- 1 scoop soy protein powder
- ¾ cup blueberries, frozen

SNACK 2

- 1 tablespoon sunflower seeds
- 1 cup fat-free cottage cheese
- ¾ cup pineapple

DAY 22

2,000 kcal

BREAKFAST

- 4 4-inch whole wheat pancakes
- ½ cup fat-free cottage cheese
- ½ mango
- 1 cup non- or low-fat plain soy milk

LUNCH

- 2 cups dark leafy greens with tomato, mushrooms, onions, shredded carrots, celery (add vegetables of your choice)
- 6 ounces flavored tofu, firm
- ⅔ cup beans
- 1 tablespoon crushed walnuts
- 1 hard-boiled egg
- *Dressing: 1 teaspoon olive oil, 1 tablespoon rice vinegar, turmeric, pepper, garlic clove, crushed

DINNER

- 8 ounces bluefish, poached with fat-free, low-sodium vegetable broth
- 6 ounces yams cut in quarters, roasted with 2 cups eggplant, cubed
- Cook potatoes and eggplant with 2 teaspoons olive oil, 1 teaspoon turmeric, garlic, ¼ cup chopped onion, sea salt and pepper to taste
- 1 cup skim milk

SNACK 1

- 1 cup Greek yogurt
- ¾ cup blueberries

SNACK 2

- 1 ounce cheese, 75% light
- 1 pear

DAY 23

2,000 kcal

BREAKFAST

- 2 eggs with ½ cup tomato and onions
- 1 ounce cheese, 75% light
- 1 slice whole wheat bread
- ¾ cup blueberries
- 1 cup non- or low-fat plain soy milk

LUNCH

- ¾ cup tuna fish made with 1½ teaspoons light mayo and turmeric to taste
- Place tuna over 1 cup mixed greens
- 10 whole grain, low-fat crackers
- 1 cup skim milk

DINNER

- 8 ounces grilled chicken
- 1 cup roasted Brussels sprouts
- ⅔ cup brown rice
- 1 teaspoon olive oil in cooking
- 4 ounces red wine or grape juice

SNACK 1

- 1 apple
- 4 teaspoons natural peanut butter

SNACK 2

- ½ cup nonfat cottage cheese
- ¾ cup raspberries
- 6 almonds

DAY 24

2,000 kcal

BREAKFAST

- ½ cup oatmeal, cooked with 1 cup skim milk
- ¾ cup blueberries
- 1 hard-boiled egg

LUNCH

- 6 ounces ham, low-sodium
- 1 ounce low-fat cheese (Alpine Lace)
- 2 slices whole wheat bread
- Mustard
- Add tomato, lettuce, sprouts to sandwiches
- 1 cup non- or low-fat plain soy milk

DINNER

- *Shrimp Sauté
- 8 ounces shrimp
- ½ cup sautéed vegetables (spinach, tomato, mushrooms, broccoli, garlic, etc.) cooked with 1 teaspoon olive oil
- 2 tablespoons Parmesan cheese
- 1 cup whole wheat pasta, cooked

SNACK 1

- 1 ounce dark chocolate, 70% cocoa
- 12 cherries

SNACK 2

- 1 cup non-fat cottage cheese
- ½ banana
- 1 tablespoon crushed walnuts

DAY 25

2,000 kcal

BREAKFAST
- 2 flax waffles
- ¾ cup blackberries
- 3 egg whites with 1½ ounces 75% light cheddar cheese
- 1½ cups nonfat plain soy milk

LUNCH
- *6 ounces grilled chicken salad, made with 1½ teaspoons light canola mayo and turmeric
- 1 slice whole wheat toast
- ½ cup raw string beans

DINNER
- *Tofu stir-fry (6 ounces firm tofu), cooked with 1 teaspoon olive oil
- ¼ pound shrimp
- 3 cups vegetables
- ⅔ cup brown rice
- ⅔ cup white beans
- 4 ounces red wine or grape juice

SNACK 1
- 4 teaspoons natural peanut butter
- 1 apple

SNACK 2
- 1 cup fat-free Greek yogurt
- ¾ ounce freeze-dried blueberries

DAY 26

2,000 kcal

BREAKFAST
- 4 4-inch whole wheat pancakes
- Pancake topping: ½ banana, mashed with 1 tablespoon crushed walnuts
- 1 hard-boiled egg
- 1 cup non- or low-fat plain soy milk

LUNCH
- 8-ounce tuna steak, grilled with soy marinade (1 teaspoon olive oil, 2 tablespoons soy sauce, 1 teaspoon turmeric)
- ⅓ cup whole wheat couscous
- 1 cup or 8 spears asparagus

DINNER
- 1 cup whole wheat pasta
- 4 soy meatballs, approx. 1 inch diameter
- 1 cup low-fat marinara sauce
- 4 ounces red wine or grape juice

SNACK 1
- 3 cups nonfat popcorn
- ¾ ounce freeze-dried blueberries
- 1 ounce cheese, 75% light
- 1 cup skim milk

SNACK 2
- 1 cup Greek yogurt
- ¾ cup blackberries
- 1 teaspoon vanilla

DAY 27

2,000 kcal

BREAKFAST
- ½ cup steel-cut oatmeal, cooked with:
- 1 cup skim milk
- 1 tablespoon crushed walnuts
- 2 hard-boiled eggs

LUNCH
- Sandwich made with 2 slices light whole grain bread and mustard
- 8 slices soy turkey
- 1 slice low-fat soy cheese
- 1 cup baby carrots, marinated in ½ cup balsamic vinegar and 1 teaspoon turmeric

DINNER
- Appetizer: 5 whole grain crackers and 2 ounces 75% light cheese
- 8 ounces salmon, baked or broiled
- 1 cup cooked kale made with 1 teaspoon sesame oil, pepper, and fresh garlic
- 1 6-ounce baked yam

SNACK 1
- *Peanut Butter Smoothie
- ½ cup nonfat plain yogurt
- ½ banana
- 2 teaspoons natural peanut butter
- 3 ice cubes

SNACK 2
- ¾ ounce freeze-dried blueberries
- 6 almonds

DAY 28

2,000 kcal

BREAKFAST
- 2 flax waffles
- 1¼ cups whole strawberries
- 3 egg whites + 1 egg with 1½ ounces cheddar cheese, 75% light
- 1 cup skim milk

LUNCH
- *Turmeric Tuna Salad
- 1 cup tuna fish with 1 tablespoon light canola mayo and turmeric
- Serve over 1 cup raw baby spinach
- 1 slice whole wheat toast

DINNER
- *4 ounces grilled tempeh, cooked with 1 teaspoon sesame oil
- 1½ cups steamed broccoli, cut into small pieces
- ⅔ cup black beans
- ½ cup barley
- Mix broccoli, beans, and barley, and add rice vinegar, sea salt, and fresh pepper to taste

SNACK 1
- 6 almonds
- 1 cup nonfat plain yogurt
- 1 peach, sliced
- Cinnamon to taste

SNACK 2
- 1 cup fat-free cottage cheese
- ¾ cup blueberries

Recipes

The Anti-Alzheimer's Diet Recipes are made with the same brain-boosting and healing foods I recommended in Step 1. Filled with antioxidants, flavonoids, folate, omega-3 fatty acids, and other key nutrients, the recipes are *just what the doctor ordered* for a healthy brain.

Whether you enjoy low-carb cuisine, a vegetarian diet, fresh fish, or meat at mealtime, you'll find something you will enjoy in these recipes. If you use the recipes to follow the 28-day Menu, you will be using ⅓ good fats, ⅓ lean protein, and ⅓ complex carbohydrates throughout the day, as I recommend for the Anti-Alzheimer's Diet.

In addition, I've given you some kids' lunch ideas that follow the foods recommended in the Anti-Alzheimer's Diet. Drop me an e-mail at our Web site—www.anti-alzheimers.com—and let me know if you and your children enjoy these foods as much we do.

Drinks

Brain-Boosting Peanut Butter
Smoothie
Brain-Boosting Berry Smoothie

Breakfast Foods

Morning Glory Cereal
Blueberry Apple Muffins
Asparagus Omelet

Snacks

Salty and Sweet Snack Pack
Chocolate Dream Snack Pack
Almond Snack Pack
Yogurt Parfait
Garlicky Thyme Cheese Spread

Entrées

Italian Style Soy Meatballs
Grilled Tempeh
Soy-ish Burgers
Mediterranean Style Cod
Shrimp and Vegetable Sauté
Scallop and Pasta Medley
Tofu Stir-Fry
Tuna on the Barbie
Chilean Sea Bass
Rosemary Chicken Salad
Rosemary Rubbed Pork
Tenderloin
Cider Chicken

Salads

Turmeric Tuna Salad
Tomato Salad
Grilled Chicken Salad
Grilled Chicken Pasta Salad
Couscous Salad
Three-Bean Salad

Sauces and Dressings

Rosemary Marinade
Lemon Sauce for Salmon
Soy Ginger Marinade
Turmeric Vegetable Glaze
Citrus Marinade
AARX Salad Dressing
Spice of Life

Vegetables

Grilled Asparagus
Sesame Kale
Roasted Carrots
Yam and Eggplant Compote
Turmeric Spaghetti Squash

School Lunch Suggestions

DRINKS

Brain-Boosting Peanut Butter Smoothie

½ cup nonfat plain yogurt
½ banana, sliced
2 teaspoons natural peanut butter with omega-3 fatty acids
  (check food label)
3 ice cubes

Blend all ingredients in a blender until smooth. More ice may
be needed depending on desired consistency.

1 serving

Smoothie provides: ~130 calories, ~14 grams carbohydrate,
5 grams fat, ~6 grams protein

Brain-Boosting Berry Smoothie

1 cup skim milk
1 scoop soy protein powder
¾ cup blueberries (or other favorite berry), frozen
3 ice cubes

Blend all ingredients in a blender until smooth. More ice may
be needed depending on desired consistency.

1 serving

Smoothie provides: ~180 calories, ~27 grams carbohydrate,
~1 gram fat, ~15 grams protein

BREAKFAST FOODS

Morning Glory Cereal

1 cup skim milk
¼ cup dry oatmeal
¾ cup blueberries
1 tablespoon crushed walnuts
Cinnamon (optional)

Bring ½ cup milk to a boil. Stir in oats and cook for about 5 minutes over medium heat. Add blueberries, walnuts, and additional milk. Cook together for 2 additional minutes. Add cinnamon if desired!

1 serving

1 serving provides: ~ 275 calories, ~42 grams carbohydrate, ~11 grams protein, ~6 grams fat

Blueberry Apple Muffins

1 cup plain nonfat yogurt
1 cup chopped apple (1 medium), remove skin
½ cup blueberries
½ cup blueberry applesauce
2 egg whites
1 cup oat bran
1 cup whole wheat flour
1 tablespoon baking powder
1 teaspoon ground cinnamon
½ teaspoon ground nutmeg
1 tablespoon packed brown sugar

Heat oven to 400°F. Spray 12 medium muffin cups with non-stick cooking spray or line with paper baking cups. Mix yogurt, apple, blueberries, blueberry applesauce, and egg whites in large bowl until blended. Stir in remaining ingredients (except brown sugar) just until flour is moistened. Fill muffin cups ⅞ full. Sprinkle with brown sugar. Bake for 20 to 24 minutes or until golden brown. Immediately remove from pan.

12 servings (12 muffins)

1 serving provides: ~85 calories, ~17 grams carbohydrate, ~4 grams protein, ~1 gram fat

Asparagus Omelet

½ cup or 4 spears asparagus, steamed (left over from dinner the night before)
1 ounce low-fat cheese (try Cabot, 75% light cheddar cheese)
2 eggs
Ground fresh pepper

Cut asparagus into bite-size pieces, shred or dice 1 ounce of cheese, and scramble eggs. Mix asparagus, egg, and fresh pepper. Coat a small nonstick pan with olive oil spray and heat over medium heat. Add egg mix and let it flatten out on the bottom of the pan. When the eggs are cooked enough to fold over (about 3 minutes), add cheese and fold egg mix into a half-moon shape. Take omelet out of the pan when the cheese is melted and the eggs are thoroughly cooked.

1 serving

1 serving provides: ~135 calories, 5 grams carbohydrate, 16 grams protein, 6 grams fat

SNACKS

Salty and Sweet Snack Pack

3 cups nonfat popcorn
¾ ounce freeze-dried blueberries
Salt (optional)

Mix all ingredients in a ziplock bag. Take this snack with you wherever you go. . . . Great snack in combination with an ounce of low-fat cheese!

1 serving

1 serving provides: ~140 calories, ~30 grams carbohydrate (one ounce of low-fat cheese will add about 55 calories, 7 grams of protein, and 3 grams of fat)

Chocolate Dream Snack Pack

½ ounce dark chocolate chips
2 tablespoons dried cranberries (no sugar added)
1 tablespoon crushed walnuts

You can double, triple, or quadruple the ingredients. Separate into equal parts and store in small plastic reusable containers for on-the-go snack packs.

1 serving

1 serving provides: ~190 calories, ~22 grams carbohydrate, 0 grams protein, ~10 grams fat

Almond Snack Pack

12 almonds

12 pieces of your favorite whole grain low-sugar cereal (e.g., Cheerios)

2 tablespoons raisins (or 2 tablespoons similar-size, no-sugar-added dried fruit)

Mix all ingredients in a plastic bag. A great snack when you're on the go!

1 serving

1 serving provides: ~150 calories, 15 grams carbohydrate, ~2 grams protein, ~10 grams fat

Yogurt Parfait

1 cup nonfat plain yogurt
Cinnamon to taste
½ banana, sliced
6 almonds

Mix yogurt and cinnamon. Refrigerate for 1 hour to let the cinnamon flavor infuse the yogurt (this step is optional). When you are ready to eat, add banana and almonds!

1 serving

1 serving provides: ~195 calories, ~27 grams carbohydrate, ~8 grams protein, ~5 grams fat

Garlicky Thyme Cheese Spread

2 tablespoons roasted garlic, mashed
1 tablespoon thyme, finely chopped
2¼ teaspoons fresh squeezed lemon juice
1 quart low-fat plain yogurt
½ teaspoon Worcestershire sauce
Salt and pepper

In a mixing bowl, add garlic, thyme, and lemon juice to the yogurt and mix well. Place mixture in cheesecloth and set in a colander over a bowl (to catch liquid). Place in the refrigerator for 12 hours. Discard liquid and place remaining yogurt in bowl. Add salt and pepper to taste. Spread on whole grain crackers or celery stalks.

4 servings

1 serving provides: ~90 calories, 12 grams carbohydrate, ~8 grams protein, 0 grams fat

ENTRÉES

Italian Style Soy Meatballs

¼ cup marinara sauce
3 tablespoons water
1 teaspoon Worcestershire sauce
⅔ cup dry textured soy protein
½ cup chopped mushrooms
½ cup chopped onion
1 cup grated zucchini
2 cloves garlic, minced

2 tablespoons chopped fresh parsley

¾ cup canned soybeans, drained and mashed

¾ teaspoon each basil and oregano

¼ teaspoon each sage and salt

⅛ teaspoon pepper

¾ cup dry bread crumbs, whole wheat

¼ cup rolled oats

Combine the marinara sauce, water, and Worcestershire sauce, and bring to a boil in a small saucepan. Pour over the textured soy protein and set aside until ready to use. Cook the mushrooms, onions, zucchini, and garlic in a nonstick pan. Add water if needed to prevent sticking, but cook away any excess liquid. Combine the textured soy protein, cooked vegetables, and remaining ingredients in a mixing bowl. Mix together well. Using 2 tablespoons of mixture at a time, form balls and place on a nonstick baking sheet. Bake for 20 minutes at 350°F, turning once. (If you don't want to bake, brown the balls in a nonstick skillet over moderate heat.) Double the recipe so that you will have meatballs left over for lunch or to freeze for use at dinner another night!

8 servings (2 meatballs per serving)

1 serving provides: ~140 calories, ~18 grams carbohydrate, ~12 g protein, ~3 grams fat

Grilled Tempeh

2 8-ounce packages tempeh

3 cloves garlic, minced

2 tablespoons finely minced ginger

1 teaspoon turmeric

¼ teaspoon red pepper flakes

4 teaspoons sesame oil
4 teaspoons brown sugar
3 tablespoons light soy sauce

Steam tempeh for 20 minutes in a covered pot. Mix all ingredients in a large bowl to create a marinade. Portion out tempeh to 4 equal portions. Marinate for 20 minutes. Grill over medium heat for about 5 minutes per side.

4 servings

1 serving provides: ~240 calories, ~20 grams carbohydrate, ~24 grams protein, ~9 grams fat

Soy-ish Burgers

1 cup textured soy protein
1 cup water
1 pound lean ground turkey
1 envelope low-sodium dry onion mix
¼ cup dried onion flakes
2 cups bread crumbs, whole wheat
1 teaspoon turmeric
1 teaspoon parsley flakes
1 teaspoon garlic powder
Ground pepper to taste

In a mixing bowl, add water to textured soy protein. Let stand 10 minutes to allow soy protein to absorb water. Mix all ingredients and form burger patties. Refrigerate for 30 minutes. Grill until burger is thoroughly cooked.

8 servings (1 burger per serving)

1 serving provides: ~250 calories, ~26 grams carbohydrate, ~27 grams protein, ~4 grams fat

Mediterranean Style Cod

1 medium onion, sliced thin
3 ounces fresh baby spinach
1 pound cod fillets .
1 small zucchini, sliced thin
4 ounces cremini mushrooms, sliced
1 14½-ounce can crushed tomatoes
1 teaspoon dried oregano
2 cloves garlic, diced
⅓ cup dry white wine
3 tablespoons fresh lemon juice
Ground pepper to taste

Preheat oven to 425°F. Place onions and spinach in the bottom of a baking dish and lay cod on top. Scatter zucchini and mushrooms over the fish. Add additional ingredients. Cover and place in oven for 15–20 minutes, until cod flakes easily with a fork.

2 servings

1 serving provides: ~350 calories, ~34 grams carbohydrate, ~48 grams protein, ~5 grams fat

Shrimp and Vegetable Sauté

Olive oil cooking spray
2 teaspoons olive oil
2 cloves garlic, chopped
8 ounces shrimp

½ cup broccoli florets
½ cup baby spinach
¼ cup chopped mushrooms
Ground fresh black pepper
Hot red pepper flakes
1 cup whole wheat pasta, cooked
2 tablespoons Parmesan cheese

Coat a nonstick pan with olive oil cooking spray and add 1 teaspoon of the olive oil. Turn on stove to medium heat and add 1 of the garlic cloves. Once the pan is heated and garlic is simmering, add the shrimp. Cook shrimp until they turn pink. Place shrimp in a separate dish, away from the heat (to prevent shrimp from becoming overcooked). Add 1 teaspoon of olive oil and cooking spray back into the pan with the left-over garlic clove. Once the garlic is slightly cooked, add the broccoli. After about 5 minutes, add the spinach and mushrooms. Once the vegetables are thoroughly cooked, add the shrimp back into the pan and cook together for 5 minutes or until the shrimp are thoroughly cooked. Place over the whole wheat pasta and add the Parmesan cheese.

1 serving

1 serving provides: ~520 calories, ~30 grams carbohydrate, ~34 grams protein, ~22 grams fat

Scallop and Pasta Medley

1 cup spinach, steamed
½ cup whole wheat pasta, cooked al dente
Olive oil cooking spray
½ tablespoon olive oil
½ tablespoon pine nuts

2 cloves garlic, minced
10 ounces scallops
2 tablespoons Parmesan cheese

Prior to cooking scallops, steam spinach in the microwave and cook the pasta, then hold both on the side. Coat a nonstick pan with olive oil cooking spray. Over low to medium heat add olive oil and pine nuts. When oil starts to simmer, add the garlic. When the pine nuts are slightly browned, add the scallops and sauté until they are thoroughly cooked. Add the spinach and whole wheat pasta. Mix all ingredients together over low heat for 2–3 minutes. Add Parmesan cheese before serving.

1 serving

1 serving provides: ~550 calories, ~25 grams carbohydrate, ~47 grams protein, ~29 grams fat

Tofu Stir-Fry

3 cups vegetables (green onions, carrots, broccoli, celery, zucchini)—choose vegetables you enjoy!
1 teaspoon olive oil
½ cup low-sodium, fat-free chicken broth
1 teaspoon cornstarch
1 teaspoon garlic powder
Ground ginger, to taste
6 ounces firm tofu, ½-inch cubes
⅔ cup brown rice
⅔ cup white beans

Wash and cut vegetables, dry thoroughly. Coat a nonstick large wok or skillet with olive oil cooking spray and heat

1 teaspoon oil until very hot but not smoking. Add vegetables and cook for 5 minutes, stirring frequently. Cover pan, lower heat to medium-high, and cook 1 minute. Uncover, stir a little, cover and cook 1½ minutes more. Remove from heat (make sure vegetables are tender before removing from heat).

In a small saucepan pour the chicken broth. Sprinkle in cornstach and add garlic and ginger. With wire whisk, stir to dissolve. Thicken over medium heat. When slightly thick, add the cubed tofu. Stir gently to continue thickening and heating through the tofu. Combine with the vegetables. Serve over cooked brown rice and beans.

2 servings

1 serving provides: ~355 calories, ~38 grams carbohydrate, ~28 grams protein, ~11 grams fat

Tuna on the Barbie

8-ounce tuna steak
Soy ginger marinade (see recipe on page 293)
1½ teaspoons sesame seeds
Ground fresh pepper to taste

Place the tuna steak in a shallow dish, and add the soy ginger marinade. Marinate both sides of the tuna steak. Let steak sit in refrigerator for 30 minutes before grilling. Do not let tuna sit in marinade too long because the soy sauce will overpower the tuna. Sprinkle sesame seeds over the tuna steak before grilling.

Let the grill get hot and then sear tuna on both sides. Once the tuna is seared, turn down the heat and cook for 4 minutes or less per side for 1-inch steaks (depending on how you like them). Steaks should be firm with pink in the middle.

1 serving

1 serving provides: ~350 calories, 0 grams carbohydrate, ~56 grams protein, ~16 grams fat

Chilean Sea Bass

2 pounds Chilean sea bass
1 tablespoon olive oil
1 pinch sea salt
Ground fresh black pepper to taste
½ teaspoon marjoram leaves
⅓ teaspoon thyme leaves
½ teaspoon garlic powder
½ teaspoon turmeric
2 bay leaves
¾ cup chopped onion
Paprika
½ cup white wine or skim milk
Lemon wedges to garnish

Preheat oven to 350°F. Wash fish, pat dry, and place in dish. Combine oil with salt, pepper, and herbs. Drizzle over fish.

Top with bay leaves and onions. Sprinkle with paprika. Pour wine or skim milk over all. Bake, uncovered, for 20 to 30 minutes or until fish flakes easily with a fork. Serve with lemon wedges.

4 servings

1 serving provides: ~260 calories, ~2 grams carbohydrate, ~28 grams protein, ~16 grams fat

Rosemary Chicken Salad

3 celery ribs, chopped
3 cups cubed cooked chicken, skinless white meat
½ cup fat-free mayonnaise
½ cup fat-free sour cream
1 teaspoon fresh rosemary
Ground fresh black pepper to taste

Combine all ingredients in a bowl until they are thoroughly mixed.

Serve on whole grain bread or over mixed greens.

4 servings

1 serving provides: ~185 calories, ~9 grams carbohydrate, ~22 grams protein, ~9 grams fat

Rosemary Rubbed Pork Tenderloin

Olive oil cooking spray
4 teaspoons dried rosemary
5 large cloves garlic, crushed
1 tablespoon cold-pressed olive oil
1 teaspoon coarse salt
Ground fresh black pepper to taste
2 ½-pound boneless pork loin roast, well trimmed

Preheat oven to 400°F. Line 13 x 9 x 2-inch roasting pan with foil and spray with olive oil cooking spray. Mix the rosemary, garlic, olive oil, and salt and pepper in bowl. Rub rosemary mixture all over pork. Place pork, fat side down, in the prepared roasting pan. Roast pork 30 minutes. Then turn roast fat side up. Roast until thermometer inserted into center of pork registers 155°F, about 25 minutes longer. Remove from oven and let stand 10 minutes.

Pour any juices from roasting pan into a small saucepan; set over low heat to keep warm. Cut pork crosswise into slices ⅓ inch thick. Arrange pork slices on platter. Pour pan juices over.

8 servings

1 serving provides: ~230 calories, 0 grams carbohydrate, ~28 grams protein, ~13 grams fat

Cider Chicken

4 thin boneless, skinless chicken breasts (about 6 ounces each)
1 tablespoon olive oil
4 cloves garlic, peeled and cut in half lengthwise
1 teaspoon kosher salt
Fresh ground black pepper to taste
½ cup apple cider vinegar
½ cup low-sodium chicken broth
2 teaspoons dried rosemary

Pound the chicken breasts between two sheets of waxed paper until they are of uniform thickness. Heat the oil and garlic in a nonstick pan over medium-high heat until the oil simmers and the garlic browns. (Don't burn the garlic.) Add the chicken breasts and cook until golden, 3 to 5 minutes on each side. Add the salt, pepper, vinegar, chicken broth, and rosemary. Cover, reduce heat to medium-low, and cook 3 minutes longer or until the chicken is fork-tender. Remove the chicken to a platter and keep warm. Increase heat to high and boil the sauce for approximately 3 to 5 minutes. Pour the sauce over the chicken.

4 servings

1 serving provides: ~243 calories, 0 grams carbohydrate, ~42 grams protein, ~10 grams fat

SALADS

Turmeric Tuna Salad

¾ cup tuna fish, packed in water
1½ teaspoons light mayo
1 teaspoon mustard
1 tablespoon fresh lemon juice
1 teaspoon turmeric
2 tablespoons chopped red onion
2 tablespoons chopped celery
Ground fresh pepper to taste

In a small mixing bowl, mix all ingredients. Place over greens or on whole wheat toast.

1 serving

1 serving provides: ~187 calories, 2 grams carbohydrate, 21 grams protein, 11 grams fat

Tomato Salad

This salad can be interchanged with a 1-cup vegetable serving in the menus provided.

1 cup cherry tomatoes, quartered
¼ cup red onion sliced thin
¼ cup red wine vinegar
4 large fresh basil leaves, chopped
1 teaspoon garlic powder
1 teaspoon turmeric
Ground sea salt to taste
Ground pepper to taste

Mix all ingredients. Let tomato salad sit at room temperature 30 minutes before serving.

1 serving

1 serving provides: ~50 calories per serving, ~10 grams carbohydrate, ~4 grams protein, 0 grams fat

Grilled Chicken Salad

5 ounces grilled chicken, diced
1 tablespoon light mayonnaise
¼ teaspoon turmeric
8 dried cranberries (no sugar added), cut in half
2 tablespoons diced red onion
2 tablespoons diced celery

Mix all ingredients together in a bowl. Refrigerate for 30 minutes before serving. Place on whole grain bread or over mixed baby greens.

1 serving

1 serving provides: ~240 calories, ~5 grams carbohydrate, ~35 grams protein, ~10 grams fat

Grilled Chicken Pasta Salad

6 ounces chicken breast, diced
¼ cup balsamic vinegar
½ cup whole wheat pasta, cooked al dente
⅔ cup beans
1½ cups cooked vegetables, your choice
2 cloves garlic, minced
2 teaspoons olive oil
1 pinch rosemary
Fresh ground pepper to taste

Marinate chicken in balsamic vinegar, grill until cooked thoroughly, and dice. Cook pasta and hold to the side. Open a can of beans, drain and rinse (to remove excess sodium) and hold to the side. Sauté your choice of veggies (green, yellow, and red peppers) and the garlic in a nonstick pan coated with olive oil cooking spray and 2 teaspoons olive oil. When the vegetables are cooked, add pasta, chicken, rosemary, and pepper (if desired). Optional: add additional balsamic vinegar to taste. Stir salad and serve.

1 serving

1 serving provides: ~595 calories, ~45 grams carbohydrate, ~61 grams protein, ~33 grams fat

Couscous Salad

2 cups water
¼ teaspoon turmeric powder
¼ teaspoon cinnamon
¼ teaspoon chopped fresh gingerroot
2 cups whole wheat couscous

1 cup chopped zucchini
½ cup chopped carrot
½ cup chopped green onion
¼ cup chopped green bell pepper
¼ cup chopped red bell pepper
15-ounce can garbanzo beans, drained and rinsed
2 tablespoons cold-pressed olive oil
2 tablespoons lemon juice
Salt to taste
¼ cup sliced almonds

In a 1-quart pot, bring water and spices to a boil. Stir in couscous. Cover pan and remove from heat; let stand for 15 minutes. Transfer to a large mixing bowl and add vegetables and garbanzo beans. In a small bowl, combine oil, lemon juice, and salt. Pour over salad and mix well, breaking up any clumps. Cover and refrigerate for 8 hours or overnight. To serve, sprinkle with almonds.

8 servings

1 serving provides: ~173 calories, ~21 grams carbohydrate, ~6 grams protein, ~6 grams fat

Three-Bean Salad

15-ounce can cannellini beans, rinsed well and drained
15-ounce can kidney beans, rinsed well and drained
15-ounce can garbanzo beans, rinsed well and drained
2 celery stalks, chopped fine
½ red onion, chopped fine
1 small orange bell pepper, chopped into small pieces
1 small green bell pepper, chopped into small pieces
1 cup finely chopped fresh flat-leaf parsley

1 tablespoon finely chopped fresh rosemary
⅓ cup apple cider vinegar
1 teaspoon turmeric
1 tablespoon granulated sugar
¼ cup olive oil
1½ teaspoons salt
Fresh ground black pepper to taste

In a large bowl, mix the beans, celery, onion, peppers, parsley, and rosemary. In a separate small bowl, whisk together the vinegar, turmeric, sugar, olive oil, salt, and pepper. Add the dressing to the beans. Toss to coat. Chill beans in the refrigerator for several hours, to allow the beans to soak up the flavor of the dressing.

8 servings

1 serving provides: ~255 calories, 33 grams carbohydrate, ~10 grams protein, ~9 grams fat

SAUCES AND DRESSINGS

Rosemary Marinade

Use as a marinade on meats or as a drizzle over greens.

1 tablespoon olive oil
¼ cup lemon juice
3 cloves garlic, pressed
1 teaspoon dried thyme
2 teaspoons dried rosemary
½ teaspoon lemon zest

Mix ingredients together in a jar. Refrigerate for several hours to allow flavors to blend before using.

2 servings

1 serving provides: ~82 calories, ~3 grams carbohydrate, 0 grams protein, ~7 grams fat

Lemon Sauce for Salmon

¼ cup fresh lemon juice
½ cup chicken broth, low-sodium and fat-free
1 tablespoon cornstarch
2 tablespoons honey
1 teaspoon turmeric
Ground pepper

Whisk lemon juice, chicken broth, and cornstarch. Heat in a small saucepan over medium heat until sauce thickens, continuously stirring. Stir in honey, turmeric, and pepper, and remove from heat. Top broiled salmon with about 3 tablespoons of lemon sauce.

4 servings

1 serving provides: ~15 calories, ~9 grams carbohydrate, ~1 gram protein, ~0 grams fat

Soy Ginger Marinade

This marinade can be used for chicken, fish, or pork.

1 teaspoon olive oil
2 tablespoons soy sauce
1 teaspoon turmeric

1 teaspoon garlic powder
Fresh ginger, grated, to taste

Mix all ingredients. Place your choice of meat in marinade for about 10 minutes before cooking.

1 serving

1 serving provides: ~45 calories, 0 grams carbohydrate, 0 grams protein, ~9 grams fat

Turmeric Vegetable Glaze

Use this glaze on about 1 pound of vegetables of your choice.

1 tablespoon olive oil
1 teaspoon turmeric
1 tablespoon fresh lemon juice
Pinch of sea salt

Cook vegetables of your choice until just tender. Drizzle ingredients over the vegetable. Mix vegetable and glaze well and serve.

1 serving

1 serving provides: ~135 calories, 0 grams carbohydrate, 0 grams protein, ~15 grams fat

Citrus Marinade

2 teaspoons garlic powder
1 teaspoon turmeric
½ cup orange juice
1 tablespoon fresh squeezed lemon juice

Mix all ingredients. Pour over meat, fish, or poultry. Marinate at least 10 minutes before cooking.

1 serving (½ cup)

1 serving provides: ~35 calories, 9 grams carbohydrate, 0 grams protein, 0 grams fat

AARX Salad Dressing

1 pinch turmeric
1 pinch rosemary
3 pinches garlic powder
Salt (optional)
Black pepper (optional)
1 tablespoon olive oil
2–3 tablespoons rice vinegar to taste

Mix all ingredients together and shake!

1 serving

1 serving provides: ~135 calories, 0 grams carbohydrate, 0 grams protein, ~15 grams fat

Spice of Life

This spice mix can be added to rice and bean dishes, eggs, hummus, and marinades.

¼ cup turmeric
2½ tablespoons ground coriander
1½ teaspoons ground ginger
½ teaspoon ground fresh pepper
½ teaspoon cayenne

½ teaspoon cinnamon
½ teaspoon ground cardamom
½ teaspoon fresh grated nutmeg
½ teaspoon garlic powder
Pinch of ground cloves

Mix all ingredients and pack in an airtight jar. Store in the refrigerator for up to 2 months.

⅓ cup

Provides minimal nutrition.

VEGETABLES

Grilled Asparagus

8 spears asparagus (equivalent to about 1 cup)
Garlic powder to taste
Ground fresh pepper to taste

Blanch the asparagus before placing on the grill: Place asparagus into boiling salt water for 1 minute, remove the asparagus, then submerge it into cold water and remove immediately to prevent further cooking.

Place asparagus on the grill perpendicular to the bars of the grate so the spears do not fall through. Leave a bit of space between asparagus stalks. Grill asparagus until crisp tender, approximately 5–6 minutes (2–3 minutes per side), turning once.

1 serving

1 serving provides: ~50 calories, 10 grams carbohydrate, 6 grams protein

Sesame Kale

Before eating or cooking kale, wash the leaves thoroughly under cool running water to remove any sand or dirt. Both the leaves and the stem of kale can be eaten.

Olive oil cooking spray
1 teaspoon sesame oil (per person)
2 cloves fresh garlic, minced
1 large bunch kale
Ground fresh pepper to taste
1 pinch salt

Coat a nonstick pan with olive oil cooking spray and add 1 teaspoon (per person) of sesame oil and garlic cloves. Heat oil and garlic until slightly simmering. Add kale and cook until leaves are tender. You may want to cover the pan so that steam helps to speed the cooking process.

1 serving (1 cup, cooked)

1 serving provides: ~95 calories, ~10 grams carbohydrate, ~4 grams protein, ~5 grams fat

Roasted Carrots

1 cup carrots
2 cloves garlic, minced
1 teaspoon rosemary, dried
1 teaspoon turmeric
1 teaspoon olive oil
Olive oil cooking spray

Preheat oven to 400°F. In a ziplock bag, add all ingredients and shake. Empty bag into a nonstick baking pan coated with the cooking spray. Place baking pan in the oven for about 30 minutes or until carrots are cooked thoroughly. You may want to broil carrots for the last 5 minutes of cooking until they are slightly browned. If you are cooking for more than one person, double the recipe.

1 serving

1 serving provides: ~95 calories, ~10 grams carbohydrate, ~4 grams protein, ~5 grams fat

Yam and Eggplant Compote

Olive oil cooking spray
6-ounce yam, cut in 1½-inch cubes
¼ cup chopped onion
2 teaspoons dried rosemary
2 cloves garlic, minced
2 cups eggplant cut in 2-inch cubes
2 teaspoons olive oil
Sea salt to taste
Ground fresh pepper to taste

Preheat oven to 375°F. Spray a nonstick baking pan with the cooking spray. Add the yam (wash thoroughly), onion, rosemary, and garlic, and cover tightly. Cook for 20 minutes. Remove cover and add eggplant, olive oil, salt, and pepper. Mix together well. Stir ingredients every 10 minutes. Roast until yam and eggplant are thoroughly cooked (about 30 minutes).

1 serving

1 serving provides: ~300 calories, ~40 grams carbohydrate, ~10 grams protein, ~10 grams fat

Turmeric Spaghetti Squash

1 spaghetti squash, cooked in microwave (see instructions below)

¼ cup water

2 tablespoons olive oil

1 teaspoon garlic powder

¾ teaspoon turmeric

⅛ teaspoon ground ginger

Sea salt to taste

Ground fresh black pepper

¼ cup toasted pine nuts

To microwave: Cut squash in half lengthwise; remove seeds. Place squash, cut sides up, in a covered microwave dish with ¼ cup water. Cook on high for 10 to 12 minutes. Add more cooking time if necessary. Let stand covered for 5 minutes. Use a fork to "comb" out the squash.

In a large skillet over medium heat, heat oil and then stir in spices, except salt and pepper. Cook for about a minute. Stir in squash and sauté until well coated. Add salt and pepper to taste. Remove to serving dish. Sprinkle with toasted pine nuts before serving. Servings vary depending on size of squash.

~6 servings

1 serving provides: ~140 calories, 10 grams carbohydrate, ~4 grams protein, ~10 grams fat

SCHOOL LUNCH SUGGESTIONS

Many schools have limited healthy lunch options for children to choose from; therefore, I believe it's imperative that children bring their lunches from home. Keep in mind that children should play a role in deciding on their lunch menu for the week. Try not to get into a lunchtime rut. . . . offer a variety of foods at lunchtime if your child is open to it!

Think about food safety and purchase an insulated lunch bag or box. Place a frozen bottle of water or an ice pack in the lunch bag to keep foods from spoiling. If the school your child attends has microwave access, you can pack leftovers for lunch. Leftover meals should include a vegetable from the night before! Otherwise, you can pack more traditional lunches. Lunches can include an entrée, fruit, and/or vegetable and a "treat" snack. See examples below:

Entrées

- Hot soup packed in a thermos
- Sandwiches
  - Choose **whole grain** breads, pitas, wraps, crackers, and mini-bagels.
  - Fill sandwiches with lean, low-sodium deli meats and low-fat cheeses that do not contain fillers or cereals, by-products, artificial flavors or colors
  - Also try sandwich fillers such as;
    - Hummus
    - Low-fat tuna (chunk light tuna and low-fat mayonnaise)
    - Egg salad (make with low-fat mayonnaise and *remove every other yolk*)
    - Chicken salad (use leftover grilled chicken and low-fat mayonnaise)
    - Veggie burgers
    - Baked tofu (flavored)

- Veggie slices (soy based), avocado, and tomato sandwiches
- Fresh mozzarella, tomato, and pesto
- Nut butters (as long as your child's school permits nuts on campus)

Snacks

Vegetables served with dip:
- Use a low-fat salad dressing such as Newman's Own Lighten Up! or make a low-sodium dip with low-fat sour cream.
- Offer a variety of vegetables, such as steamed asparagus, radishes, peppers, broccoli, cauliflower, cherry tomatoes, edamame, carrots, celery.

Fruit:
- Offer a variety of fruits, such as apple slices drizzled with lemon juice, fruit salad, melon, berries, kiwis, avocados, bananas, pears, nectarines, plums, peaches, cherries, mangos, papaya, figs.

Treats:
- Try to limit treats to between 100 and 150 calories at lunchtime. You can read Nutrition Facts labels and determine the serving size of snacks. Based on the serving size and the calories per serving, you can figure out how much of that particular snack you can pack for your child. If you don't want to figure out the servings on your own, you can purchase prepackaged 100-calorie snack packs.

- 100-calorie snack packs
- Low-fat cheese sticks
- Stonyfield Farm low-fat yogurt and smoothies
- Whole wheat pretzels
- Pepperidge Farm Goldfish

Drinks

- Water
- Skim or 1% milk
- Unflavored soy or rice milk drink boxes
- 100% fruit juice (limit fruit juice consumption to 4–6 ounces per day)

APPENDIX B

Figure Out Your Body Mass Index

The body mass index (BMI) is a ratio of your height and weight and can often determine if an individual is overweight or obese. To calculate your BMI, use the formula on page 304. Then using the chart on page 304, you can see that if your BMI is between 25 and 30, you are considered overweight; if it is above 30, you are considered obese. For instance, if you are 5' 5" tall and weigh 151 pounds, you are considered to be "overweight." If you are the same height and weigh 180 pounds, you are in the "obese" category. If you are 5' 11", you are overweight at 179 pounds and obese at 215 pounds. Again, as I have discussed, being overweight or obese is associated with increased levels of pro-inflammatory markers and an increased chance of Alzheimer's disease.

While the BMI is used by many health care practitioners, it can be misleading in some adults, especially men and women who have large frames or an abundance of muscle. Even though these men and women may have a BMI over 25, they might not have excessive body fat.

BMI	WEIGHT STATUS
Below 18.5	Underweight
18.6-24.9	Normal
25.0–29.9	Overweight
30.0 and Above	Obese

BODY MASS INDEX

	Normal						Overweight					Obese										Extreme Obesity														
BMI	19	20	21	22	23	24	25	26	27	28	29	30	31	32	33	34	35	36	37	38	39	40	41	42	43	44	45	46	47	48	49	50	51	52	53	54
Height (inches)												Body Weight (pounds)																								
58	91	96	100	105	110	115	119	124	129	134	138	143	148	153	158	162	167	172	177	181	186	191	196	201	205	210	215	220	224	229	234	239	244	248	253	258
59	94	99	104	109	114	119	124	128	133	138	143	148	153	158	163	168	173	178	183	188	193	198	203	208	212	217	222	227	232	237	242	247	252	257	262	267
60	97	102	107	112	118	123	128	133	138	143	148	153	158	163	168	174	179	184	189	194	199	204	209	215	220	225	230	235	240	245	250	255	261	266	271	276
61	100	106	111	116	122	127	132	137	143	148	153	158	164	169	174	180	185	190	195	201	206	211	217	222	227	232	238	243	248	254	259	264	269	275	280	285
62	104	109	115	120	126	131	136	142	147	153	158	164	169	175	180	186	191	196	202	207	213	218	224	229	235	240	246	251	256	262	267	273	278	284	289	295
63	107	113	118	124	130	135	141	146	152	158	163	169	175	180	186	191	197	203	208	214	220	225	231	237	242	248	254	259	265	270	278	282	287	293	299	304
64	110	116	122	128	134	140	145	151	157	163	169	174	180	186	192	197	204	209	215	221	227	232	238	244	250	256	262	267	273	279	285	291	296	302	308	314
65	114	120	126	132	138	144	150	156	162	168	174	180	186	192	198	204	210	216	222	228	234	240	246	252	258	264	270	276	282	288	294	300	306	312	318	324
66	118	124	130	136	142	148	155	161	167	173	179	186	192	198	204	210	216	223	229	235	241	247	253	260	266	272	278	284	291	297	303	309	315	322	328	334
67	121	127	134	140	146	153	159	166	172	178	185	191	198	204	211	217	223	230	236	242	249	255	261	268	274	280	287	293	299	306	312	319	325	331	338	344
68	125	131	138	144	151	158	164	171	177	184	190	197	203	210	216	223	230	236	243	249	256	262	269	276	282	289	295	302	308	315	322	328	335	341	348	354
69	128	135	142	149	155	162	169	176	182	189	196	203	209	216	223	230	236	243	250	257	263	270	277	284	291	297	304	311	318	324	331	338	345	351	358	365
70	132	139	146	153	160	167	174	181	188	195	202	209	216	222	229	236	243	250	257	264	271	278	285	292	299	306	313	320	327	334	341	348	355	362	369	376
71	136	143	150	157	165	172	179	186	193	200	208	215	222	229	236	243	250	257	265	272	279	286	293	301	308	315	322	329	338	343	351	358	365	372	379	386
72	140	147	154	162	169	177	184	191	199	206	213	221	228	235	242	250	258	265	272	279	287	294	302	309	316	324	331	338	346	353	361	368	375	383	390	397
73	144	151	159	166	174	182	189	197	204	212	219	227	235	242	250	257	265	272	280	288	295	302	310	318	325	333	340	348	355	363	371	378	386	393	401	408
74	148	155	163	171	179	186	194	202	210	218	225	233	241	249	256	264	272	280	287	295	303	311	319	326	334	342	350	358	365	373	381	389	396	404	412	420
75	152	160	168	176	184	192	200	208	216	224	232	240	248	256	264	272	279	287	295	303	311	319	327	335	343	351	359	367	375	383	391	399	407	415	423	431
76	156	164	172	180	189	197	205	213	221	230	238	246	254	263	271	279	287	295	304	312	320	328	336	344	353	361	369	377	385	394	402	410	418	426	435	443

Source: Adapted from Clinical Guidelines on the Identification, Evaluation, and Treatment of Overweight and Obesity in Adults: The Evidence Report.
Free access: http://www.nhlbi.nih.gov/guidelines/obesity/bmi_tbl.htm

Your Personal Waist-to-Hip Worksheet

To determine your waist-to-hip ratio, measure your waist in inches at its narrowest point, and then divide this number by your hip measurement in inches at the widest point.

Waist (in inches) _____ divided by Hip (in inches) _____ = _____

Here's a way to figure out whether you need to do something about your weight. You measure your "fat point." Grab your measuring tape, and follow these steps:

1. Measure your waist just above the navel.
2. Measure your hips at the widest part.
3. Divide your waist measurement by the hip measurement.

For example, if your waist is 30 and your hips are 40, the ratio will be 0.75. A safe waist-to-hip ratio is anything less than 0.8.

For women the waist-to-hip score should not exceed *0.80.*

For men the score should not exceed *0.95.*

APPENDIX C

Are you fit? Take these three tests that need no equipment to let you know what condition you are in: poor, average, or very good. According to the results, you should modify the exercise programs given.

1. Maximum Push-ups Test

Preparation: Lie face down.
Men: Toes on the floor, knees off the ground, hands wider than shoulders, and body straight. Start with arms extended and go down till chest just touches the floor and then up till arms are fully extended.

Men

Age	Poor	Average	Very good
30–39	3–7	14–25	35+
40–49	2–5	11–21	31+
50–59	1–4	9–18	26+
60–69	1–2	6–16	24+

Women: Toes on floor and knees touching the floor. Then do as above.

Women

Age	Poor	Average	Very good
30–39	1–4	11–22	29+
40–49	1–3	9–18	25+
50–59	1–2	7–14	21+
60–69	1	5–12	18+

2. One-minute Crunches Test

Lie face up on floor and do as many crunches as possible in one minute.
Preparation: Face up, keep knees flexed at right angle. Flex spine by squeezing abs while pushing lower back on floor. Extend fingertips to touch the top of your knees.

Men

Age	Poor	Average	Very good
30–39	18–22	27–29	35–41
40–49	13–17	22–24	29–35
50–59	9–12	17–20	25–31
60–69	7–10	14–18	22–28

Women

Age	Poor	Average	Very good
30–39	1–4	11–22	29–35
40–49	1–3	9–18	25–32
50–59	1–2	7–14	21–25
60–69	1	5–12	18–23

3. Maximum Squats Test

This test is simple; all you need is a chair.

Preparation: With a chair behind you and with feet shoulder width apart, stand as if you were about to sit on the chair. Put hands on hips. Squat till your buttocks touch the seat of the chair and then stand up again. Repeat. Remember to keep knees above feet throughout motion.

Men

Age	Poor	Average	Very good
36–45	20–23	28–31	37–42
46–55	13–17	23–27	32–36
56–65	10–13	19–22	27–32
65+	8–11	17–20	25–29

Women

Age	Poor	Average	Very good
36–45	10–15	20–23	29–35
46–55	7–11	16–18	24–29
56–65	4–7	11–14	20–25
65+	3–5	9–12	18–24

REFERENCES AND
SUPPORTING RESEARCH

CHAPTER 1

Wear and tear of stress: the psychoneurobiology of aging. American Psychological Association, August 11, 2006. http://www.apa.org/releases/aging_stress06.html. Accessed August 30, 2007.

CHAPTER 2

Yaffe K, Kanaya A, Lindquist K, et al. The metabolic syndrome, inflammation, and risk of cognitive decline. *Journal of the American Medical Association.* 2004;292:2237.

Cournot M, Marquie JC, Ansiau D, et al. Relation between body mass index and cognitive function in healthy middle-aged men and women. *Neurology.* 2006;67:1208.

Razay G, Vreugdenhil A, Wilcock G. FN The metabolic syndrome and Alzheimer disease. *Archives of Neurology.* 2007 Jan;64(1):93–6.

National Center for Health Statistics. Third National Health and Nutrition Examination Survey 1988–1994. Atlanta, GA: Centers for Disease Control and Prevention; 1996.

Trayhurn P, Wood I. Adipokines: inflammation and the pleiotropic role of white adipose tissue. *British Journal of Nutrition.* 2004;92:347–55.

Whitmer RA, Gunderson EP, Barrett-Connor E, et al. Obesity in middle age and future risk of dementia: a 27 year longitudinal population based study. *BMJ.* 2005;330:1360.

Rosengren A, Skoog I, Gustafson D, Wilhelmsen L. Body mass index, other cardiovascular risk factors, and hospitalization for dementia. *Archives of Internal Medicine.* 2005;165:321.

Luchsinger JA, Tang MX, Shea S, Mayeux R. Antioxidant vitamin intake and risk of Alzheimer disease. *Archives of Neurology.* 2003;60:203.

Elias PK, Elias MF, Robbins MA, Budge MM. Blood pressure-related cognitive decline: does age make a difference? *Hypertension.* 2004;44:631.

Staessen JA, Birkenhager WH. Cognitive impairment and blood pressure: quo usque tandem abutere patientia nostra? *Hypertension.* 2004;44:612.

Slooter AJ, Cruts M, Hofman A, et al. The impact of APOE on myocardial infarction, stroke, and dementia: the Rotterdam Study. *Neurology.* 2004;62:1196.

Arvanitakis Z, Wilson RS, Bienias JL, et al. Diabetes mellitus and risk of Alzheimer disease and decline in cognitive function. *Archives of Neurology.* 2004;61:661

Biessels GJ, Staekenborg S, Brunner E, et al. Risk of dementia in diabetes mellitus: a systematic review. *Lancet Neurology.* 2006;5:64.

Geroldi C, Frisoni GB, Paolisso G, et al. Insulin resistance in cognitive impairment: the InCHIANTI study. *Archives of Neurology.* 2005;62:1067.

Forette F, Seux ML, Staessen JA, et al. The prevention of dementia with antihypertensive treatment: new evidence from the systolic hypertension in Europe (SYST-EUR) study. *Archives of Internal Medicine.* 2002;162:2046.

Cardiovascular risk factors in midlife strongly linked to risk of dementia. American Academy of Neurology, January 2005. http://www.aan.com/press/index.cfm?fuseaction=release.view&release=245. Accessed August 30, 2007.

Hayden MR, Tyagi SC. Homocysteine and reactive oxygen species in metabolic syndrome, type 2 diabetes mellitus, and atheroscleropathy: the pleiotropic effects of folate supplementation. *Nutrition Journal.* 2004;3:4.

The American Heritage Dictionary of the English Language, 4th ed. Boston, MA; Houghton Mifflin Company.

den Heijer T, Vermeer SE, Clarke R, Oudkerk M, Koudstaal PJ, et al. Homocysteine and brain atrophy on MRI of non-demented elderly. *Brain.* 2003 Jan;126 (Pt 1):170–5.

Smith AD. Homocysteine, B vitamins, and cognitive deficit in the elderly. *American Journal of Clinical Nutrition.* 2002;75:785–6.

Garcia A, Zanibbi K. Homocysteine and cognitive function in elderly people. *Canadian Medical Association Journal.* 2004;171:897.

Shcherbatykh I, Carpenter D. Free radicals and cell signaling in Alzheimer's disease. *Journal of Alzheimer's Disease.* 2007;2(11):191–205.

CHAPTER 3

Dai Q, Borenstein AR, Wu Y, et al. Fruit and vegetable juices and Alzheimer's disease: the Kame Project. *American Journal of Medicine.* 2006;119:751.

Schaefer EJ, Bongard V, Beiser AS, et al. Plasma phosphatidylcholine docosahexaenoic acid content and risk of dementia and Alzheimer disease: the Framingham heart study. *Archives of Neurology.* 2006;63:1545.

Larson EB, et al. Exercise is associated with reduced risk for incident dementia among persons 65 years of age and older. *Annals of Internal Medicine.* 2006;144:73–81.

Scarmeas N, Stern Y, Tang MX, et al. Mediterranean diet and risk for Alzheimer's disease. *Annals of Neurology.* 2006;59:912.

Fratiglioni L, Paillard-Borg S, Winblad B. An active and socially integrated lifestyle in late life might protect against dementia. *Lancet Neurology.* 2004 June;3(6):343–53.

CHAPTER 4

Lee L, Kang SA, Lee HO, et al. Relationships between dietary intake and cognitive function level in Korean elderly people. *Public Health.* 2001;115:133.

Scarmeas N, Stern Y, Tang MX, et al. Mediterranean diet and risk for Alzheimer's disease. *Annals of Neurology.* 2006;59:912.

Grundman M. Vitamin E and Alzheimer disease: the basis for additional clinical trials. *American Journal of Clinical Nutrition.* 2000;71:630S.

den Heijer T, et al. Homocysteine and brain atrophy on MRI of non-demented elderly. *Brain.* 2003;126(Pt 1):170–5.

Durga J, van Boxtel MP, Schouten EG, Kok FJ, Jolles J. Effect of 3-year folic acid supplementation on cognitive function in older adults in the FACIT trial: a randomized, double blind controlled trial. *Lancet.* 2007;369(9557):208–16.

Sano M, Ernesto C, Thomas RG, et al. A controlled trial of selegiline, alpha-tocopherol, or both as treatment for Alzheimer's disease. The Alzheimer's Disease Cooperative Study. *New England Journal of Medicine.* 1997;336:1216.

Morris MC, Evans DA, Tangney CC, Bienias JL, Wilson RS. Associations of vegetable and fruit consumption with age-related cognitive change. *Neurology.* 2006 Oct 24;67(8):1370–6.

Dai Q, Borenstein AR, Wu Y, Jackson JC, Larson EB. Fruit and vegetable juices and Alzheimer's disease: the Kame Project. *American Journal of Medicine.* 2006 Sep;119(9):751–9.

Morris MC, Evans DA, Bienias JL, et al. Dietary fat intake and 6-year cognitive change in an older biracial community population. *Neurology.* 2004;62:1573.

Morris MC, Evans DA, Tangney CC, et al. Fish consumption and cognitive decline with age in a large community study. *Archives of Neurology.* 2005;62:1849.

Chapter 5

Abbott RD, White LR, Ross GW, et al. Walking and dementia in physically capable elderly men. *Journal of the American Medical Association.* 2004;292:1447.

Weuve J, Kang JH, Manson JE, et al. Physical activity, including walking, and cognitive function in older women. *Journal of the American Medical Association.* 2004;292:1454.

Fratiglioni L, Paillard-Borg S, Winblad B. An active and socially integrated lifestyle in late life might protect against dementia. *Lancet Neurology.* 2004;3:343.

Netz Y. The effect of a single aerobic training session on cognitive flexibility in late middle-aged adults. *International Journal of Sports Medicine.* 2007 Jan;28(1):82–7.

Luchsinger JA, Reitz C, Honig LS, et al. Aggregation of vascular risk factors and risk of incident Alzheimer disease. *Neurology.* 2005;65:545.

Chapter 6

Carnero C. Education provides cognitive reserve in cognitive deterioration and dementia. *Neurologia.* 2007 Mar;22(2):78–85.

House J. Social isolation kills but how and why? *Psychosomatic Medicine.* 2001;57B(3):p 212–22.

Thompson RG, Moulin CJ, Hayre S, Jones RW. Music enhances category fluency in healthy older adults and Alzheimer's disease patients. *Experimental Aging Research.* 2005 Jan–Mar;31(1):91–9.

Fratiglioni L, Paillard-Borg S, Winblad B. An active and socially integrated lifestyle in late life might protect against dementia. *Lancet Neurology.* 2004 June;3(6):343–53.

Roe CM. Education and Alzheimer disease without dementia: support for the cognitive reserve hypothesis. *Neurology.* 2007 Jan 16;68(3):223–8.

De la Fuenta-Fernandez R. Impact of neuroprotection on incidence of Alzheimer's disease. *PLoS One.* 2006 Dec 20;1:e52.

Coyle JT. Use it or lose it—do effortful mental activities protect against dementia? *New England Journal of Medicine.* 2003;348:2489.

CHAPTER 7

Orodenker S. Family caregiving in a changing society: the effects of employment on caregiver stress. *Family and Community Health.* 1990; 12:58–70.

Christie C, Mitchell S, Bruce D. *Eat to Stay Young.* New York: Kensington; 1999.

Baltes PB, Baltes MM. Psychological perspectives on successful aging: the model of selective optimization with compensation. In: Baltes PB, Baltes MM, eds. *Successful Aging: Perspectives from the Behavioral Sciences.* New York: Cambridge University Press; 1990:1–34.

Mealer M, Shelton A, Berg B, Rothbaum B. Increased prevalence of post-traumatic stress disorder symptoms in critical care nurses. *American Journal of Respiratory and Critical Care Medicine.* 2007;175:693–97

Mattson ME. Background and rationale for the COMBINE study. Program and abstracts of the American Psychological Association 114th Annual Convention. New Orleans, LA; August 10–13, 2006.

Epel ES, et al. Stress and body shape: stress-induced cortisol secretion is consistently greater among women with central fat. *Psychosomatic Medicine.* 2000 Sept–Oct;62(5):623–32.

CHAPTER 8

Schulman KA, Berlin JA, Harless W, Kerner JF, et al. The effect of race and sex on physicians' recommendations for cardiac catheterization. *New England Journal of Medicine.* 1999;340(8): 618–26.

ACKNOWLEDGMENTS

There are numerous people I must acknowledge who have affected my life and helped with the completion of this book. To the following people, I am grateful:

My wife, Gayl, a woman among women, who as my psychotherapist kept me sane, as my editor made my copy readable, as my critic kept the book interesting, and as my wife kept me happy. Thank you for all the sacrifice you made while I was preoccupied writing *The Anti-Alzheimer's Prescription.*

My family: My mom, who at ninety-five continues to be the finest example any child could have. Not only is she patient, devoted, and loving, she had confidence in my abilities even when I didn't. My sister Elaine, whose love and devotion sustain me. My big sister, Joan, forever supportive, my mentor in Latin, French, and English. My brother, Michael, and his staff, David Genten and Ken Mendel, for their expert research on exercise and also their help in developing a video on isometrics and super-slow strengthening. My oldest son, Vinnie, who keeps me sane by being my friend and supporting this book and my golf game. My son Michael, for his artistic talent second to none. My daughter, Kaycee, who never gives up and whose kindness helps others. My son Kevin, for his computer genius and insight. My nieces, Geralyn and Joannie, for their love and feedback. My nephew, Mark, for his expertise in physical therapy; Michele, for her typing. Andy Imbus and his parents, who are like family and have been supportive of my book. My in-laws, Faye and George, who encouraged and supported my manuscript from day one.

My colleagues: Dr. Bill and Martha Sears, authors of thirty-two parenting books, for their confidence in me and their recommendation of my new literary agent, Denise Marcil. Bill gave me the ideas of the Harmonic Symphony and how bad habits that begin in childhood, including a sedentary lifestyle and poor eating, are learned from parents. Phil Schwartz, the finest neurobiologist and Christian I know. Phil's review of my work was so important as our knowledge of neurobiology grows daily. Dr. Richard Shubin, my colleague and partner, who was the instigator of this book. Richard's knowledge keeps me current and his humor makes each day of practice easier. Dr. Jay Rho, whose knowledge of neurobiology was of invaluable help. Hyman Gross, my best friend during my neurology residency, for introducing me to my wife. Drs. Bob and Rob Watkins, the finest surgeons to the major leaguers of every sport from baseball to wrestling, thanks for your help on building brawn. My other neurological partners, Ken Wogenson, Arthur An, Jessie Lu, and Morris Powazek, who help me by taking care of my patients. Drs. Bill Preston and Bruce Clearman, both fine neurologists and friends. Rex Green, the most courageous man I know, whose example kept me going on this book.

My friends Father Tad and Dr. John Haas for their support, knowledge, and ethical expertise. Lu Cortese and EWTN for their confidence in my ability to foster the word. Msgr. Gus Moretti for his prayers. The men and women of Legatus and Knights of Malta who have helped me make this a better world. Reverend Rick Warren and his *Purpose Driven Life*, a book for us all.

Dr. Marshall Wills, whose 100th birthday and tutorship the past thirty years has kept me on the straight and narrow path and is the best example of the 4-Step Method.

My office staff and research team, whose thoroughness and hard work supplied me with all the material I needed: Laura Kennedy and Chris Martinez who have stayed true to me over the years. She will make an outstanding physician someday; he a CEO. Shana, my physician's assistant, a woman who is mature beyond her years and whose help every day made the book all it could be. Giovanni, Francesco Dandekar, Olivia, and Jasmine for invaluable help down the stretch. Jo Stella and

Melodie Atkinson, my business manager and secretary. Faithful friends—more than employees—Susie, Anna, and Suzzette whose photocopying and emergency help saved the day at times.

Lauren Garguila Brand, R.D., who is like a niece to me. Lauren, the daughter of my best man, Bob, helped with the 28-day menu and recipes, and her expertise as a gifted dietician will help millions save their brains.

Dr. Joe Connors at Pasadena City College, a leader of our young students, who gives hope through his unselfish dedication and warnings to others about the dangers of Alzheimer's.

Tommy Lasorda, Mr. Baseball, the only man I've known whose love and dedication and selflessness outweigh his fame. No one can appreciate how much you do for the love of your country and its national pastime.

Bob Bancroft, who set up my Web page and who assists with the supplement BriteShield.

Tony Lasos, Jeff Drier, Didi, Patrick, and Judy and Michail Stahl for their review of the manuscript and advice. (Jeff, sorry you couldn't lose weight.)

Al and Cathy Robinson, whose friendship and marketing expertise have been invaluable.

Elizabeth George, the award-winning mystery writer, who selflessly lent me her office on the due date for submission of the book.

Dr. Mac Holder, Dr. Mike Habib, and Stan Alexander, my truest friends, who gave me an example of what it is like to be a compassionate doctor.

Dr. Ed Todd, the father of stereotactic neurosurgery, who gave me my start in writing. At eighty-seven, Ed is living proof that our minds are interest-bearing accounts. A renaissance man, Ed is a neurosurgeon and attorney and has a Ph.D. in history.

No acknowledgment is complete without mentioning one's agent, Denise Marcil. Without degrees in psychology and law, she has been my counselor and has given me counsel. I could never have accomplished this work without her. And I can never thank her enough.

My editors, Lauren Marino and Brianne Ramagosa, at Gotham

Books. Thank you for your insights and faith in my work. What amazes me most is the real modus operandi of Denise, Lauren, and Brianne, that is, the compassion they have for humanity. You supported the work because of your belief that the 4-Step Method will help save a generation from this impending epidemic of Alzheimer's. There is hope for the world with leaders like you!

Dr. Helen Cui, head of the department of neurology, for her help and knowledge.

James Russell, M.S., an illustrator, graphic designer, and photographer living near Atlanta, Georgia, whose brilliant artistic talents brought the book to a new level of excellence.

Last, but certainly not least, thank you to my medical writer, Debra Fulghum Bruce, Ph.D., from Atlanta, Georgia. With Dr. Bruce on my development and writing team, it was like having Babe Ruth, Barry Bonds, and Lou Gehrig in the batting order. How could anyone fail a writing project with you making the calls, Deb?

For all of these and so many more. . . . I am grateful and humbled.

For more information, please visit:
www.anti-alzheimers.com.

The Web site features:

- Diet tips
- Exercise instructions and videos
- Neurobics
- Relaxation techniques
- Advice on caring for Alzheimer's patients
- Clinical trials
- And more . . .

INDEX

Note: Page numbers in *italics* refer to illustrations or text boxes.

abdominal breathing, 201–2
abdominal fat, 37–38, 40, 78, 100, 179
Academy of Neurology, 110, 230
acetylcholine, *14*, 107, 147, 164
acetylcholinesterase inhibitors, 240–42
activities of daily living, 14–15, 23, 24
addictions, 71, 182, 186, 187
adrenaline
    and cortisol, 180
    and exercise/physical activity, 191
    and hormonal balance, *65*, 66, 68
    and laughter, 201
    and meditation, 197
    release of, *14*
    and stress response, 178–79, 181
age and aging
    and brain changes, 149–50
    and cortisol, 180–81
    early onset of Alzheimer's, 26–27, 187
    as risk factor, 31, 32–35, 58
    and sleep, 188
    and stress, 177
    and telemerase, 33–34
    theories of aging, 33
    and weight, *103*
    and wisdom, 155
aggressive behavior, 18, 52, 237
agnosia, 153, *153*
alcohol, 187
aldosterone, 44, *65*, 66
alprazolam (Xanax), 246

aluminum, 55
Alzheimer, Alois, 13
Ambien (zolpidem), 246
American Academy of Neurology, xi, 234
American Diabetes Association, 47
American Dietetic Association, 101
American Heart Association, 53, 88
American Psychological Association (APA),
    177
American Psychosomatic Society, 87
American Stress Institute, 181
amitriptyline (Elavil), 243, 247
amyloid plaques, 13, *14*, 67, 80, 248
amyloid protein, 67, 180, 191
anabolic hormones
    and catabolic hormones, 35, 71, 119
    and cortisol, 188
    effects of, 116
    and exercise, 48–49, 114, 129
    and forgiveness, 196
    and free radicals, 84
    in the hormonal symphony, *65*, 66–67
    and stress response, 190
    and telomeres, 34
anger, 52
anomia, 174
anthocyanins, 103
anthocyanosides, *60*
antidepressants, 203, 242–44, *244*, 246–47
antineuronal antibodies, 220
antioxidants, 42, 66, 79, 83–89, *86*, 105

antiphospholipid antibodies, 220
anxiety, 67, 71, 121, 181, 191
APOL e2 (apolipoprotein-e2) gene, 33
APOL e4 (apolipoprotein-e4) gene, 33
appetite, 182–83, 185, 186
apples, 103–4
apraxia, 153
Aricept (donepezil), 236, 240, 241
ascorbic acid, 84
aspirin, 108, 110–11, 235, 236
associative patterns, 157
atherosclerosis (hardening) of the arteries,
    78
Ativan (lorazepam), 246
attention, 164, 180
autonomic system, 118–19, 120–21, 180
axons, *14*
axon terminal buttons, *14*

Babinski's reflex test, 218
baby boomers, 10, 80, 123, 124, 148, 192–
    93, *251*
Baltimore Longitudinal Study of Aging,
    105, 121
beans, 105
beef, *89*
behavior, 14, 19–24
belly fat, 37–38, 40, 78, 100, 179
benzodiazepines, 246
berries, *60*, 103
beta-amyloid, 47, *60*, 78, 87, 248–49
beta-amyloid immunotherapy, 249
beta-carotene, 84, 106, 107
beta-secretase, 249
beverages, 102
Big Brain Academy, 162, 169
bipolar disorder, 224
blind taste tests, 171
blocking (brain freeze), 67
blood pressure
    and music, 201
    and prescription steroids, 187
    as risk factor, 55, *60*, 102
    and stress response, 118
    See also hypertension
blood sugar levels, 42, 54, 81, 82, *101*
blueberries, 84

body mass index (BMI), 40, 41–42, 49,
    303–5
brain
    anatomy of, 150–55, *151*
    atrophy of, 156
    building reserves in, 162–63
    deterioration of, 154–55, 163, 227
    and growth in adult years, 162
    making new connections in, 161–62
    maturation of, 148–50
    memory creation, 146–48
    unfit brains, 160–61
    See also cognitive skills and ability;
        limbic (emotional) brain; neocortex
        (thinking) brain
Brain Age, 157, 162, 169
brain-derived trophic factor (BDTF), 49
brain freeze (blocking), 67, 214
Brand, Lauren, 103
BriteShield, 108
Buddha study, 194
bupropion (Wellbutrin), 243, 244
B vitamins, 51, 90, 99, 219. *See also specific
    B vitamins, including* folic acid

caffeine, 187–88, 235
calcium, 95, 96, 106, 220
cancer, 40, 96, 179
Capgras Syndrome, 208
carbohydrates, 77–79, 81–83, 93, 100,
    186, 187
cardiac system, 121
cardiovascular risk factors, *47–48*
carotenoids, 84
catabolic hormones
    and anabolic hormones, 35, 71, 119
    and cortisol, 188
    effects of, 117
    in the hormonal symphony, *65*, 66–67,
        71
    and hypertension, 44
    and physical activity, 48
catechins, 85
causes of Alzheimer's, xii, 225–26. *See also*
    risk factors
cerebral cortex, 151, 152–53
ceruloplasmin cortisol, 220

checkbook balancing exercise, 170
chess, 171
children and teens, *47*, 48, 102, 111, *145*, 200
children of Alzheimer's patients, 23, 232
chocolate, 255
cholesterol, *45–46*
  and beef, *89*
  and fish and fish oils, 87
  and food choices, 94
  and metabolic syndrome, 53
  risks associated with, 96
  and simple carbohydrates, 78
  and smoking, 55
  See also HDL cholesterol; LDL cholesterol; triglycerides
choline, 107
chondroitin sulfate, 110
chromium picolinate, 108
chronic obstructive pulmonary disease (COPD), 225
chronic stress, 179–81. *See also* stress
Coenzyme Q10, *108*, 108
coffee, 187
cognitive skills and ability
  and diabetes, 46–47
  and engagement or novelty, 158–59
  and exercise, 113, 114–15
  and homocysteine, *60*, 219
  and hormonal balance, 119, 121
  and hypertension, 43
  mental agility, 148, 150
  mental capacity, 148–49
  and nutrition, 84–85
  and stages of Alzheimer's, 14, 24
  and stress, 177
  See also brain
Columbia study, 79–80
Complete Blood Counts (CBC), 218–21
computerized tomography (CT), 222
concentration, 113, 168, 180
confabulations, 238–39
control, sense of, 191, 192
coping skills and styles, 177–82, 191–92
copper, 55, 107, 110, 220
corn syrup, 97
cortisol

  and aging, *103*, 180–81
  and C-reactive proteins, 52
  effects of, 117
  and exercise/physical activity, 48, 49, 113
  and fat distribution, 179, 180
  and hormonal balance, *65*, 66, 68, 188
  and hypertension, 44
  and laughter, 201
  and meditation, 197
  and prescription steroids, 186–87
  production of, 66, 181
  role of, 178
  and sleep, 44, 181
counseling, 195
Cousins, Norman, 200
C-reactive proteins, 51–53, 102, 116, 185, 219
curcumin, 107
cure for Alzheimer's, 8, 59
Cymbalta (duloxetine), 243
cytokines, 52, 66, 188

daidzein, 105
dairy products, 81, 85, 94, 99, 105, 254
dancing, 166
dehydroepiandrosterone, 110
delusions, 237–39
dementia
  described, 13, 15
  incidence of, xii
  and social network, 160
  treatment of, 210, 223–27
dendrites, *14*, 162
denial, 59
depression
  and C-reactive proteins, 52
  and diagnosing Alzheimer's, 18, 224, 225
  and dopamine, 182, 185
  and exercise, 113
  and folate, 91
  and hormonal balance, 67, 121
  and stress, 177
Desyrel (trazodone), 243, 247
Dexamethasone, 186
diabetes
  and catabolic hormones, 117
  and C-reactive proteins, 53

diabetes (*continued*)
and exercise/physical activity, 48, 141
and glycemic index, 81
and memory performance, 82
and metabolic syndrome, 53, 54
and nutrition, 80
and obesity, 40
and prescription steroids, 187
and Real Brain Age, 58
as risk factor, 31, 46–48
and simple carbohydrates, 78
and sleep, 183
and stress, 177, 179, 190
and sugar substitutes, 98
and visceral fat, 100–103
diagnosing Alzheimer's, 207–30
and advocates for patients, 216–17
and causes of Alzheimer's, 225–26
early diagnosis, 210, 230
lab tests, 218–21
selecting a physician, 209–15, *212*, 223
steps to, 217–18
tests for, 217, 218
and treatable dementia, 223–27
diapers, 23
diazepam (Valium), 246
dictionary words exercise, 166–67
Diet, Anti-Alzheimer's
Golden Dozen food choices, 103–6
harmonic method of eating, 92–93
kitchen overhaul, 92, 93–98
menus, 257–70
recipes, 271–302
shopping strategies, 92, 96, 99, 253–56
See also nutrition
disorientation, 24
docosahexaenoic acid (DHA), *60*, 86–87
donepezil (Aricept), 236, 240, 241
dopamine
and addiction, 186
and anabolic hormones, 116
and exercise, 113
and forgiveness, 196
and hormonal balance, 10, *65*, 66, 71
and prescription steroids, 186–87
and relaxation, 181, 194, 200

release of, *14*
and sleep cycles, 182, 185, 187–88
and stress, 187–88
and weight, 40, 41
doxepin (Sinequan), 243, 246
drawing exercises, 168
driving skills, 22–23
Duke University study, 52
duloxetine (Cymbalta), 243

early onset of Alzheimer's, 26–27, 187
education, 54
Effexor (venlafaxine), 244
eggs, 99, 107
eicosanoids, 87–88
eicosapentaenoic acid (EPA), 86
Elan Pharmaceuticals, 248
Elavil (amitriptyline), 243, 247
electroencephalography (EEG), 221, *222*
electromyograms (EMG), 221, *222*
electrooculograms (EOG), 221, *222*
emergency skills, 22
emotion, 146–47, 153–54, 159, 160
endorphins, 114, 198, 201
enemas, 110
engagement, 158–59
enkephalins, 198
Environmental Protection Agency (EPA), 88
ephedra, 110
estrogen, 49, *65*, 66, 114, 116, 129, 190
eszopiclone (Lunesta), 246
evening-primrose oil, 235
executive function, 180, 186
Exelon (rivastigmine), 240, 241
exercise and physical activity, 112–45
aerobic exercise, 126–27, *139*, *140–41*, 144, 157–58
and aging, 122, 160–61
benefits of, 113–15, *115*, 122–23, *123*
and brain health, 156
caloric expenditure, *140–41*
circuit training, 139–40
and C-reactive proteins, 51
dancing, 166
and dieting, 125
and families, 102

and hormones, 10, 71, 116–19, 119–22, *123*, 129
implementing a fitness program, 141–44
importance of, 9, 201
and metabolic syndrome, 54
moderate levels of, 122–23, *127–28*
recommendations for, 48
and relaxation mode, 66
and stages of Alzheimer's, 24
strength training, 128–35, *139*, 140–41, 307–9
and stress response, 191
stretching and flexibility, 135–38
tips for, 124–25
warming up, 125
water exercise, 144
extrinsic risk factors, 31, *32*
abnormal lipids, 45–46
C-reactive proteins, 51–53
diabetes, 46–48
homocysteine, 50–51
hypertension, 43–44
metabolic syndrome (insulin resistance), 53–54
obesity, 40–43
sedentary lifestyles, 48–49
eye-hand coordination, 168

family and friends, 17, 24, 55, 160, 208, 232–33. *See also* social networks and interaction
family history, 35–36, 214, 217
fat, body, 66, 78, 179, 180
fat, dietary
and Anti-Alzheimer's Diet, 79, 80–81, 92, 93, 100
and food choices, 96
and food labels, 94–95
and glycemic index, 81
and insulin-glucagon balance, 77
fatigue, 38, 78, 121, 180
fiber, 96, *101*, 106
fibrillary tangles, 13, *14*, 52, 67, 87, 148
fibromyalgia, 67
fight or flight response, 65, 68, 178–79, 189–90

fish and fish oils
benefits of, 87
Golden Dozen food choices, 99, 104
and omega-3 and -6 fatty acids, *60*, 86, *89*, 95
and pollutants, 88, *89*
supplements, 107, *108*, 109
flavonoids, 84, 85
flaxseed and flaxseed oil, 107, 110
flours, 80, 92, 93, 97
fluvoxamine maleate (Luvox), 243
folate
and Anti-Alzheimer's Diet, 79
and Complete Blood Counts (CBC), 219
and Golden Dozen food choices, 105, 106
and homocysteine, *60*
sources of, 51, 90, *92*, 105, 106
folic acid, 51, *60*, 90–93, 107, *108*, 219
Folic Acid and Carotid Intima-media Thickness (FACIT) trial, 90–91
Food and Drug Administration (FDA), 88
forebrain, 150, *151*
foreign films, 171
foreign languages, 173
forgiveness, 196
Framingham study, *60*, 87
free radicals, 33, *60*, 66, 67, 80, 83–84
frontal inferior temporal gyrus, 194
frontal lobe, 155
fruits
and antioxidants, 85
benefits of, *60*
and fiber, *101*
and glycemic index, 81
good choices for, 99, 253
and homocysteine, 51
promoting consumption of, 102
Functional Magnetic Resonance Imaging (FMRI), 222–23

Gage, Fred, 113
galantamine (Razadyne/ER), 240, 242
gamma-secretase, 249
garlic, 107
gastric bypass surgery, 226–27

genders
and body fat, *41*
and diagnosing Alzheimer's, 228–29
and incidence of Alzheimer's, 32
and incidence of dementia, xii
and muscle mass, 114
and physical activity, 49
as risk factor, 31
and sleep issues, 183–84, 185
General Adaptation Syndrome, 178, 189
genetic predisposition for Alzheimer's
and age, 26, 32
and incidence of Alzheimer's, 26, 188
as risk factor, 31, 58
steps to counteract, *37*
testing for, 33
genistein, 105
Geodon, 237
geriatricians, *212*
ghrelin, 49, 183, 185
ginger, 110
ginkgo biloba, 110, 235, 236
Glabella Tap reflex test, 18, 218
glucagons, 77–78, 82
glucose
and catabolic hormones, 117
and cortisol, 178
and exercise/physical activity, 113, 191
and glycemic index, 82–83
and the hormonal symphony, 66, 67
and insulin, 77–78, 180
and metabolic syndrome, 54
and nutrition, 80
as risk factor, 31
role of, 178
glutamate, 147
glycemic index, 81–83
glycoproteins, 178
Golden Dozen food choices, 99, 103–6
golf, 167
grape juice, *60*, 255
grape seed extract, 108
"grasp" reflex, 18, 218
greens, 105
green tea and green tea extract, 85, 108, 235, 255

habits, 69
Halcion (triazolam), 246
hallucinations, 21, 237–39
haloperidol (Haldol), 237, 245
hardening of the arteries, 78
harmonic method of eating, 92–93, *99*, 100–103
HDL cholesterol
and metabolic syndrome, 53
and physical activity, 48
as risk factor, 31, 45, *46*
and simple carbohydrates, 78
and smoking, 55
health maintenance organizations (HMOs), 214
heart attacks, 43, 50, 185
heart disease, 40, 53, 96
herbs and spices, 107, 110, 171, 255
high-fructose corn syrup, 97
hindbrain, 150, *151*
hippocampus
and aging, 151
and brain freeze (blocking), 67
and cognitive performance, 180
deterioration of, 43–44, 155
effect of Alzheimer's on, 18
and exercise, 49, 113
and homocysteine, 90
and imaging, 222
and memory, 153–54, 160
and parasympathetic nervous system, 190
role of, *120*, 153–54
and sleep apnea, 227
and stress, 180
hip/waist ratio, 43. *See also* body mass index (BMI)
homocysteine
and cognitive skills and ability, *60*, 219
and Complete Blood Counts (CBC), 219
and C-reactive proteins, 53
and folic acid, 90
and nutrition, 91
as risk factor, 50–51
Honolulu-Asia Aging Study, 114–15
hope, 251–52

Hope College in Michigan, 196
hormones
    and exercise, 71, 116–19, *123*, 129
    and forgiveness, 196
    hormonal symphony, 64–67, *65*, 68–69
    imbalance in, 119–22
    and obesity, 41
    and psychological responses, 188–93
    and sleep, 121, 182, 185–86, 187–88
    and stress, 180, 187–88, 188–89
    and willpower, 70–71
    See also specific hormones
hostility, 52
housekeeping tasks, 23
human growth hormone
    and exercise, 49, 114, 116, 129
    in the hormonal symphony, *65*, 66
    and laughter, 201
    and sleep, 39, 183–84
    and stress response, 190
human immunodeficiency virus (HIV),
    220
humor, 200–201
Hydergine, 236
hygiene and grooming, 16, 23
hyperinsulinemia, 47
Hypertension, 197
hypertension
    and catabolic hormones, 117
    and C-reactive proteins, 51, 53
    and food choices, 96
    and hormonal balance, 121
    and metabolic syndrome, 53
    and nutrition, 80
    and obesity, 40
    and Real Brain Age, 58
    as risk factor, 31, 43–44, *47–48*, 55, *60*,
        102
    and sleep apnea, 185
    and stress, 177, 190
    and visceral fat, 100
    See also blood pressure
hypothalamus, 41, *103*, 179, 180–81

illusions, 237–39
imaging tests, *222–23*
immune system, 177, 188, 190, 197

incidence of Alzheimer's, 26, 32
incidence of dementia, xii
incontinence, 24, 239
independent functioning, 17, 26
inflammation
    and fish and fish oils, 87
    and free radicals, 83
    and obesity, 41
    and omega-3 fatty acids, 86
    pro-inflammatory markers, 40, 41, 52,
        116, 219
    and visceral fat, 100
    and vitamin C, 85
insomnia, 67, 183–84, 240, 246–47. *See
        also* sleep
Institute of Medicine, 91
insulin
    and amyloid protein, 180, 191
    and anabolic hormones, 116
    and carbohydrates, 187
    and exercise, 113, 129
    and glucagon, 77–78, 82
    and glycemic index, 81
    and the hormonal symphony, *65*, 66–
        67, 69
    and nutrition, 80, *101*
    and physical activity, 49
    and prescription steroids, 187
insulin-degrading enzyme (IDE), 67, 78
insulin-like growth factors (IGFs), 114,
        116–19
insulin resistance
    and cortisol, 179
    and dehydroepiandrosterone, 110
    and glycemic index, 81
    and hyperinsulinemia, 47
    and nutrition, 80
    and obesity, 40, 42
    and physical activity, 48
    as risk factor, 30–31, 53–54, 102
    and simple carbohydrates, 78
    and sleep, 183
    and visceral fat, 100
insurance, 214, *251*
interests, loss of, 17
intrinsic risk factors, 31, *32*, 32–37, *37*
iron, 55, 95, 96, 107

irrational thinking, 19–24
Italian Longitudinal Study on Aging, 108

Journal of Gerontology, 115
Journal of the American Medical Association, 184
judgment, 19–24

Kaufman, Fran, *47*
kava, 236

Lancet, 160
language skills, 18–19, 23, 153, 169
Lasorda, Tommy, 75–76
LDL cholesterol
    and excess glucose, 66
    and flavonoids, 85
    as risk factor, 45, *45*, 102
    and simple carbohydrates, 78
    and smoking, 55
learning, 17–19, 54, 113, 163
legumes, 81, 99, *101*, 105, 254
leptin, 49, *103*, 183, 185
life balance, 192–93
life expectancies, xi
lifestyle and Alzheimer's, 36, *60*, 175, 188, 226
limbic (emotional) brain
    and autonomic system, 118–19, 120–21
    components of, *120*
    deterioration of, 155
    and hormonal balance, 10, 68, 71, 187–88
    location of, *151*
    and neocortex (thinking) brain, 69, 70, 150
lipids, 31, 45–46, *47–48*, 98. *See also* cholesterol
list memorization exercise, 165
L-lipoic acid, 108
locus coeruleus, 152
long-term potentiation, 156
lorazepam (Ativan), 246
lumbar punctures, 220
Lunesta (eszopiclone), 246
lymphocytes, 180

macroglia, 67
magnesium, 220
magnetic resonance imaging (MRI), 222
math exercises, 170
mealtime exercises, 171
medications, xii, 8, 203, 215, 233–34, 240–47
meditation, 194, 197–98
Mediterranean diet, *60*, 79–80
Medrol, 186
melatonin, 110
memantine HCL (Namenda), 242
memory
    creation of memories, 146–48
    and deterioration of brain, 155, 163
    and diabetes, 82
    and emotion, 146–47, 153–54, 160
    and exercise, 113
    exercises (mental) for, 164–67, 170
    and folic acid, 90–91
    loss of, 13, 250
    and mild cognitive impairment (MCI), 16
    and olfactory cells and connections, 159
    and sleep cycles, 182
    and stages of Alzheimer's, 17–19, 24
mental agility, 148, 150
mental capacity, 148–49
menus, 100–101, 257–70
metabolic syndrome (insulin resistance)
    and cortisol, 179
    and dehydroepiandrosterone, 110
    and glycemic index, 81
    and hyperinsulinemia, 47
    and nutrition, 80
    and obesity, 40, 42
    and physical activity, 48
    as risk factor, 30–31, 53–54, 102
    and simple carbohydrates, 78
    and sleep, 183
    and visceral fat, 100
metabolism, 114, 121, 161, 183
microglia, 52
midbrain, 150, *151*
middle age, 26–27
mild cognitive impairment (MCI), 15–17, 19, 157, 210

milk, 85, 255
Mind Fit, 162
mirtazapine (Remeron), 244
mitochondria, 67
mobility, 24
monoclonal antibody treatments, 248–49
monounsaturated fats, 80, 95, 104
motor functions, 23
MSG (monosodium glutamate), 98
multisensory associations, 163
multivitamins, 107, *108*, 110
muscle pain, 67
muscles mass, 24, 49, *103*, 121, 180
music, 201

Namenda (memantine HCL), 242
names, remembering, 161–62
National Heart, Lung, and Blood Institute, 53
National Institute on Aging, 183
National Institutes of Health (NIH), 80, 84
neocortex (thinking) brain
    deterioration of, 152, 155
    and hormonal balance, 9–10, 71
    and limbic (emotional) brain, 69, 70, 150
    location of, *151*
    role of, 152–53
nervous system
    parasympathetic nervous system, 118–19, 190
    sympathetic nervous system, 65, *65*, 66, 117, 178, 180, 189–90
    vagal nervous system, *65*, 66, 116, 181
neurasiums, 157
neurobics (brain exercises), 9, 146–73
    engagement in life, 158–59, 172–73
    implementation of, 163–73
    importance of, 160–61
    mealtime exercises, 172
    memorization exercises, 164–67
    and neural connections, 161–62
    and olfaction, 159
    purpose of, 156, 157–58
    reasoning exercises, 171
    thinking exercises, 169–70
    visualization exercises, 167–69

neurogenesis, 49, 113, 162
neuroleptics, 245
neurologists, *212*
neuronal forest, *14*, 147–48
neurons, *14*, 156, 161
neuroplasticity, 49, 162, 164
neuropsychologists, *212*
neurotransmitters
    and exercise, 113
    in the hormonal symphony, 65, *65*, 66
    and neuronal forest, *14*, 147
    and obesity, 41
    and sleep, 182
    and stress, 180
    See also specific neurotransmitters
neurotrophins, 158
New England Journal of Medicine, 229
nonsteroid anti-inflammatory drugs (NSAIDs), 110–11
noradrenaline, *65*, 66, 180, 181
norepinephrine, 52, 152
nortriptyline (Pamelor), 243
novelty, 159
number memorization exercise, 164–65
Nurses' Health Study, 115
nursing homes, 26, 27, 234, 240, *251*
nutrition, 75–111
    and antioxidants, 83–85
    and children, *47*, 111
    and fiber, *101*
    and folic acid, 90–93
    and food labels, 93–96
    and glycemic index, 81–82
    Golden Dozen food choices, 99, 103–6
    and "good fats," 80–81 (*see also* fat, dietary)
    and the hormonal symphony, 71
    and insulin-glucagon balance, 77–78
    and Mediterranean diet, *60*, 79–80
    and omega-3 fatty acids, 86–89
    and processed foods, 92, 93, 96–97
    role in Alzheimer's, 42, *60*
    and stages of Alzheimer's, 24
    and supplements, 107–11, *108*, 215, 235
    and vitamins, 51
    See also Diet, Anti-Alzheimer's
nuts, 81, 99, *101*, 104, 253

obesity
  and C-reactive proteins, 51, 53
  and metabolic syndrome, 53
  and nutrition, 79
  and physical activity, 48
  and Real Brain Age, 58
  as risk factor, 31, 40–43
  and sleep loss, 182–83, 186
  and weight loss, 100–102
olanzapine (Zyprexa), 237, 245–46
Olestra, 110
olfactory cells and connections, 17, 18,
  159, 171
omega-3 fatty acids, *60*, 79, 80, 86–89, *89*,
  94–95, 104
omega-6 fatty acids, 94–95
oxazepam (Serax), 246
oxidative stress, *60*

Palmar Mental Reflex test, 18, 218
Pamelor (nortriptyline), 243
paranoid ideation, 17, 208
parasympathetic nervous system, 118–19,
  190
Parkinson's disease, 152, 159
paroxetine (Paxil), 243
partially hydrogenated fats, 93, 98
passionflower, 235
Paxil (paroxetine), 243
pedometers, 126–27
perception of stressors, 189–91
personality types, 192
personal skills, 155
pets, 197
phosphorus, 106, 220
physicians, 209–15, *212*, 223, *251*
phytochemicals, 104
plaques, 52, 67, 87, 148, 248–49
polycystic ovarian syndrome (PCOD), 53
polysomnography (sleep studies), 221, *222*
polyunsaturated fats, 80–81, 94
Positron Emission Tomography (PET),
  222–23
potassium, 106
Prednisone, 186–87
prescription steroids, 186
primary care physicians, *212*, 214

processed foods, 92, 93, 96–97
progesterone, 49, 117, 190
prognosis after initial diagnosis, 24, 234
pro-inflammatory markers, 40, 41, 52, 116,
  219
prostaglandins, 52
Prostate-Specific Antigen (PSA), 225
proteins, 77, 79, 81–82, 93, 99, 100, 254
Prozac, 236, 243
psychiatrists, *212*
psychological responses, 188–93
psychoneuroimmunology, 38
psychosis, 237–39

quercetin, 103–4
quetiapine (Seroquel), 237, 245

Radica Brain Games, 162
ramelteon (Rozerem), 246
Razadyne/ER (galantamine), 240, 242
reaction times, 16
reading, 164
Reagan, Ronald, 39
Real Brain Age, 57–71
  and hormonal symphony, 64–67, *65*,
    70–71
  quiz, 61–64
  and willpower, 67–69, 70–71
reasoning exercises, 171
recipes, 271–302
recognition of family and friends, 24
red wine, *60*, 108, 255
reflex tests, 18, 217, 218
relaxation, 181, 193–94, 198–200
religion, 194, 196
Remeron (mirtazapine), 244
rest and recovery, 193–203
  exercise and physical activity, 201
  identification and elimination of
    stressors, 195
  laughter, 200–201
  medications, 203
  meditation, 197–98
  music, 201
  relaxation, 198–200
  sleep, 202
  social support, 197

spiritual or religious pursuits, 196
talking with someone, 195
time for oneself, 196
Restoril (temazepam), 246
resveratrol, 108
retirement, 172–73, *251*
rivastigmine (Exelon), 240, 241
rigidity, 24
risk factors, 30–56, *32*, 225–26
    extrinsic risk factors, 31, *32*, 40–54
    intrinsic risk factors, 31, *32*, 32–37, *37*
    other risk factors, 54–56
    sentinel risk factors, 31, *32*, 37–39, 58,
        113, 177, 203
Risperdal, 237
road trips, exercises for, 166, 167
Rotterdam Study, 43–44, 85, 87
route memorization exercise, 166
Rozerem (ramelteon), 246

Saint John's-wort, 110, 236
salt, 98, 187–88
saturated fats, 42, *89*, 93, 94, 96, 98
Schulman study, 229
Sears, William, 111, *145*, 173
secretase inhibitor research, 249
sedentary lifestyles, 9, 48–49, 157
seeds, 81, 99, *101*
selective serotonin reuptake inhibitors
    (SSRIs), 203, 243–44
selenium, 84, 108
Selye, Hans, 178
sensorimotor speed, 90–91
sentinel risk factors, 31, *32*, 37–39, 58,
    113, 177, 203
Serax (oxazepam), 246
Seroquel (quetiapine), 237, 245
serotonin
    and anabolic hormones, 116
    and antidepressants, 203
    and exercise, 113
    and forgiveness, 196
    and hormonal balance, 10, *65*, 66, 71
    and prescription steroids, 186–87
    and relaxation, 181, 194
    release of, *14*
    and sleep, 182, 186, 187–88

and stress, 187–88
and weight, 40, 41
sertraline (Zoloft), 243
service groups, 173
serving sizes, 93–94
Shakespeare, William, 169
shopping strategies, 92, 96, 99, 253–56
Simon game, 168
Sinequan (doxepin), 243, 246
sleep
    and aging, 188
    and cortisol, 44, 181
    and C-reactive proteins, 53
    disturbances in, 23
    and exercise, 113
    and hormonal balance, 10, 121, 182,
        185–86
    and human growth hormone, 39, 183–84
    importance of, 182–86
    and insomnia, 67, 183–84, 240, 246–47
    medications for, 187, *203*, 240, 246–47,
        247
    and prescription steroids, 186
    and relaxation, 198
    as risk factor, 9, 31, 38–39, 58
    sleep apnea, 39, 184–85, 226–27
    strategies for achieving, 202
    and stress, 180, 181, 187–88
    and weight, 182–83
Sleep, 182–83
smell, sense of, 17, 18, 159, 171
smoking, *47–48*, 51, 55
snoring, 39, 184
Snout reflex test, 18, 218
Snyder, Evan, 156
social inhibitions, 21–22
social networks and interaction, 55, *60*,
    156, 160, 197, *251*
sodium, 92, 96, 98
solitaire, 168
Sonata (zaleplon), 246
soy products, 99, 105, 254
specialists, 223
spices and herbs, 107, 110, 171, 255
stages of Alzheimer's
    mild cognitive impairment (MCI), 15–
        17, 19, 157, 210

stages of Alzheimer's (*continued*)
stage 1: mild impairment, 17–19, *25*
stage 2: moderate impairment, 19–24, *25*
stage 3: severe impairment, 24, *25*, 153
starchy foods, 79, 81, 255
stem cells, 8, 156, 162, 248, 250
stepping program, 126–27, 144, 157–58
stress
and aging, 177
and coping skills, 177–82
and C-reactive proteins, 53
described, 177
and exercise/physical activity, 48, 113–14
and hormonal balance, 10, 188–89
and hypertension, 44
and locus coeruleus, 152
perception of stressors, 189–91
and prescription steroids, 186–87
as risk factor, 9, 31, 37–38, 58
signs and symptoms of, 181–82
and sleep, 180, 181, 187–88
and telemerase, 33–34
and theories of aging, 33
uncontrolled chronic stress, 179–81, 189
strokes, 43, 185
substantia nigra, 152
sugars, 80, 92, 93, 97, 178
sugar substitutes, *98*, 98
sundowning, 21–22
supplements, 107–11, *108*, 215, 235
sweet potatoes, 106
sympathetic nervous system, *65*, 65, 66, 117, 178, 180, 189–90
symptoms of Alzheimer's, *25*, 237–40. *See also* stages of Alzheimer's

tangles, 13, *14*, 52, 67, 87, 148
tau, 87, 249
teas, 85, 235, 255
telemerase and aging, 33–34
television, 157
telomeres, 179
temazepam (Restoril), 246
testosterone, 49, *65*, 66, 114, 116, 129, 190

therapies for Alzheimer's, 231–52
future treatment options, 247–50
holistic remedies, 235–36
medications, 8, 233–34, 236, 240–47
supplements and home remedies, 235
treating symptoms, 237–40
thinking exercises, 169–70
Thorazine, 237
thyroid, 66, *103*, 114, 116, 129
thyroid panels, 219–20
"tip of the tongue inability," 18–19
tomatoes, 106
trans fats, 42, *89*, 93, 94, 96, 98
transitional state, 15–17
trazedone (Desyrel), 243, 247
triazolam (Halcion), 246
tricyclics, 243
triglycerides
and fish and fish oils, 87
and metabolic syndrome, 53
as risk factor, 31, 45
role of, 178
sources of, 78, 98
trophic hormones, 49
turmeric, 107, 108
tyrosine, 105

uncontrolled chronic stress, 179–81, 189
University of Pittsburgh, 184
University of Wisconsin, 194
U.S. Department of Agriculture (USDA), 80
U.S. Department of Health and Human Services, 48

vaccines for Alzheimer's, 248
vagal nervous system, *65*, 66, 116, 181
valerian, 236
Valium (diazepam), 246
vascular dementia, 47, *60*, 210
vegetables
and antioxidants, 85
benefits of, *60*
and cognitive performance, 84
and fiber, *101*
and glycemic index, 81
good choices for, 99, 104, 253–54

and homocysteine, 51
promoting consumption of, 102
venlafaxine (Effexor), 244
video games, 157, 162
violence, 18, 218
visualization exercises, 167–69
vitamin A, 95, 96
vitamin B$_6$, 51, 106, 219, 226–27
vitamin B$_{12}$, 51, 210, 219, 223–24, 226–27
vitamin C, 84, 85, 95, 96, 106, 107
vitamin E, 42, 84, 85, 107
VO2 MAX Step Test, 142
vocabulary words exercise, 166–67
volunteering, 172–73, 196

waist circumference, 31, 37–38, 43, 49, 53, 78
wandering, 240
water exercise, 144

weight loss, 100–102, *101*, 110–11, 181, 185, 226–27
Wellbutrin (bupropion), 243, 244
Wells, Marshall, 149
whole grains, 81, 93, 99, *101*, 106
willpower, 69, 70–71
wisdom, 155
women. *See* genders
Women's Health Study, 82
"worried well," 227–28

Xanax (alprazolam), 246

zaleplon (Sonata), 246
zinc, 55, 220
Zoloft (sertraline), 243
zolpidem (Ambien), 246
Zutphen Elderly Study, 87
Zyprexa (olanzapine), 237, 245–46